140
Structure and Bonding

Series Editor: D. M. P. Mingos

For further volumes:
http://www.springer.com/series/430

Structure and Bonding

Series Editor: D. M. P. Mingos

Recently Published and Forthcoming Volumes

Zintl Ions
Principles and Recent Developments
Volume Editor: Thomas F. Fässler
Vol. 140, 2011

Zintl Phases
Principles and Recent Developments
Volume Editor: Thomas F. Fässler
Vol. 139, 2011

Inorganic 3D Structures
Volume Editor: Angel Vegas
Vol. 138, 2011

Molecular Catalysis of Rare-Earth Elements
Volume Editor: Peter W. Roesky
Vol. 137, 2010

Metal-Metal Bonding
Volume Editor: Gerard Parkin
Vol. 136, 2010

Functional Phthalocyanine Molecular Materials
Volume Editor: Jianzhuang Jiang
Vol. 135, 2010

Data Mining in Crystallography
Volume Editors: Hofmann, D. W. M., Kuleshova, L. N.
Vol. 134, 2010

Controlled Assembly and Modification of Inorganic Systems
Volume Editor: Wu, X.- T.
Vol. 133, 2009

Molecular Networks
Volume Editor: Hosseini, M. W.
Vol. 132, 2009

Molecular Thermodynamics of Complex Systems
Volume Editors: Lu, X., Hu, Y.
Vol. 131, 2009

Contemporary Metal Boron Chemistry I
Volume Editors: Marder, T. B., Lin, Z.
Vol. 130, 2008

Recognition of Anions
Volume Editor: Vilar, R.
Vol. 129, 2008

Liquid Crystalline Functional Assemblies and Their Supramolecular Structures
Volume Editor: Kato, T.
Vol. 128, 2008

Organometallic and Coordination Chemistry of the Actinides
Volume Editor: Albrecht-Schmitt, T. E.
Vol. 127, 2008

Halogen Bonding
Fundamentals and Applications
Volume Editors: Metrangolo, P., Resnati, G.
Vol. 126, 2008

High Energy Density Materials
Volume Editor: Klapötke, T. H.
Vol. 125, 2007

Ferro- and Antiferroelectricity
Volume Editors: Dalal, N. S., Bussmann-Holder, A.
Vol. 124, 2007

Photofunctional Transition Metal Complexes
Volume Editor: V. W.W. Yam
Vol. 123, 2007

Single-Molecule Magnets and Related Phenomena
Volume Editor: Winpenny, R.
Vol. 122, 2006

Zintl Ions

Principles and Recent Developments

Volume Editor: Thomas F. Fässler

With contributions by

G.S. Armatas · B. Eichhorn · T.F. Fässler · S. Gärtner ·
M.G. Kanatzidis · R.B. King · S. Kocak · N. Korber

Springer

Editor
Thomas F. Fässler
Department of Chemistry
Technical University Munich
Lichtenbergstraße 4
85747 Garching
Germany
thomas.faessler@lrz.tum.de

ISSN 0081-5993 e-ISSN 1616-8550
ISBN 978-3-642-26875-5 ISBN 978-3-642-21181-2 (eBook)
DOI 10.1007/978-3-642-21181-2
Springer Heidelberg Dordrecht London New York

Softcover reprint of the hardcover 1st edition 2011

Cover design: eStudio Calamar, Berlin/Figueres, Spain

Printed on acid-free paper

Springer is part of Springer Science+Business Media (www.springer.com)

Series Editor

Prof. D. Michael P. Mingos

Principal
St. Edmund Hall
Oxford OX1 4AR, UK
michael.mingos@st-edmund-hall.oxford.ac.uk

Volume Editor

Thomas F. Fässler

Department of Chemistry
Technical University Munich
Lichtenbergstraße 4
85747 Garching
Germany
thomas.faessler@lrz.tum.de

Editorial Board

Structure and Bonding Also Available Electronically

Structure and Bonding is included in Springer's eBook package *Chemistry and Materials Science*. If a library does not opt for the whole package the book series may be bought on a subscription basis. Also, all back volumes are available electronically.

For all customers who have a standing order to the print version of *Structure and Bonding*, we offer the electronic version via SpringerLink free of charge.

If you do not have access, you can still view the table of contents of each volume and the abstract of each article by going to the SpringerLink homepage, clicking on "Chemistry and Materials Science," under Subject Collection, then "Book Series," under Content Type and finally by selecting *Structure and Bonding*.

You will find information about the

- Editorial Board
- Aims and Scope
- Instructions for Authors
- Sample Contribution

at springer.com using the search function by typing in *Structure and Bonding*.

Color figures are published in full color in the electronic version on SpringerLink.

Aims and Scope

The series *Structure and Bonding* publishes critical reviews on topics of research concerned with chemical structure and bonding. The scope of the series spans the entire Periodic Table and addresses structure and bonding issues associated with all of the elements. It also focuses attention on new and developing areas of modern structural and theoretical chemistry such as nanostructures, molecular electronics, designed molecular solids, surfaces, metal clusters and supramolecular structures. Physical and spectroscopic techniques used to determine, examine and model structures fall within the purview of *Structure and Bonding* to the extent that the focus is on the scientific results obtained and not on specialist information concerning the techniques themselves. Issues associated with the development of bonding models and generalizations that illuminate the reactivity pathways and rates of chemical processes are also relevant.

The individual volumes in the series are thematic. The goal of each volume is to give the reader, whether at a university or in industry, a comprehensive overview of an area where new insights are emerging that are of interest to a larger scientific audience. Thus each review within the volume critically surveys one aspect of that topic and places it within the context of the volume as a whole. The most significant developments of the last 5 to 10 years should be presented using selected examples to illustrate the principles discussed. A description of the physical basis of the experimental techniques that have been used to provide the primary data may also be appropriate, if it has not been covered in detail elsewhere. The coverage need not be exhaustive in data, but should rather be conceptual, concentrating on the new principles being developed that will allow the reader, who is not a specialist in the area covered, to understand the data presented. Discussion of possible future research directions in the area is welcomed.

Review articles for the individual volumes are invited by the volume editors.

In references *Structure and Bonding* is abbreviated *Struct Bond* and is cited as a journal.

Impact Factor in 2009: 4.152; Section "Chemistry, Inorganic & Nuclear":
Rank 2 of 43; Section "Chemistry, Physical": Rank 7 of 113

Preface

Clusters that exclusively consist of metal atoms often are rather elusive and thus have mostly been detected and investigated in the gas phase or are the playground for theoreticians. However, well-defined bare metal clusters of p block elements can be obtained in macroscopic scale from *Zintl phases*, which have been discovered, studied and named after the German chemist *Eduard Zintl*. Such phases can be described as intermetallic compounds in which the valence electrons of the more electropositive component are transferred to the more electronegative partner under formation of salt-like compounds which in many cases contain anionic cluster units of p block metals and semi-metals as building blocks. This subject is treated in a preceding book of this series in Volume 139 (*Zintl Phases*). In some cases, *Zintl phases* are soluble in polar aprotic solvents, and the solutions contain intact moderately charged, discrete so-called *Zintl anions*, which then can be used in situ as starting materials for the synthesis of a plethora of main group element clusters. Prominent examples for such 'extractable' polyanions are ${E^{V}}_7^{3-}$ and ${E^{IV}}_9^{4-}$, which some of them have been observed in solution already at the end of the nineteenth century. However, the exploration of their fascinating chemistry dates back only a few decades.

This book documents the metamorphosis of *Zintl ions* from objects of virtu to valuable precursors for the synthesis of larger main group element cluster units of well-defined composition and size. Historical milestones of *Zintl ion* chemistry have been reviewed (Eduard Zintl: His life and scholar work. Kniep R in [1]; [2, 3]), but are also mentioned in the introductory remarks of some of the contributions to this book. The chemical bonding in homo- and heteroatomic Zintl ions and their structures are summarized and compared to those of related main group element clusters, with main focus on deltahedral polyanions of Group 14 elements in the chapter by *RB King*. *N Korber* and *S Gärtner* describe polyanions of Group 14 and 15 elements in alkali and alkaline earth metal solid state compounds and solvates in a subsequent chapter. The Chapter by *B Eichhorn* and *S Kocak* is dedicated to the dynamic behavior of Group 14 element Zintl ions and their derivatives in solution which has extensively been studied by NMR experiments. The rich chemistry of Group 14 and Group 15 element polyanions with special emphasis on the formation of intermetalloid clusters ('molecular alloys') and their relationship to ligand-stabilized cage molecules is outlined by *TF Fässler*, and the last chapter by *M Kanatzidis* and *GS Armatas* focuses on the use of Zintl ions for the synthesis of Ge-based materials for possible photovoltaic applications.

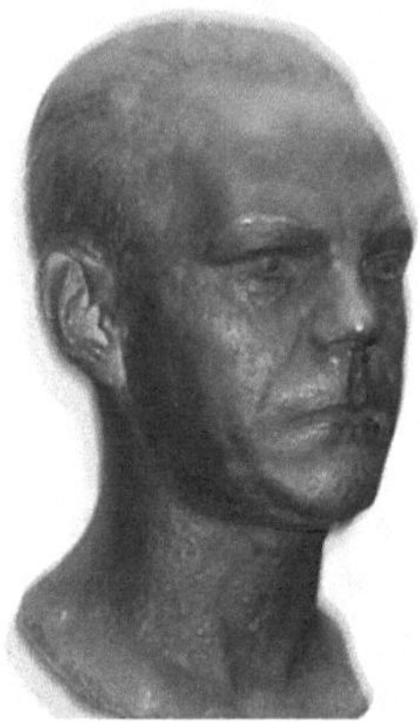

Bust of Eduard Zintl at the Technical University of Darmstadt

[1] Kauzlarich SM (ed) (1996) Chemistry, structure, and bonding of zintl phases and ions. VCH, Weinheim
[2] Corbett JD (1985) Polyatomic Zintl anions of the post-transition elements. Chem Rev 85: 383–397
[3] Fässler TF (2001) The renaissance of homoatomic nine-atom polyhedra of the heavier carbon-group elements Si to Pb. Coord Chem Rev 215:347–377

Technical University Munich
April 2011

T.F. Fässler

Contents

Struct Bond (2011) 140: 1–24
DOI: 10.1007/430_2010_21

Published online: 12 May 2011

Structure and Bonding in Zintl Ions and Related Main Group Element Clusters

R.B. King

Abstract This chapter reviews the methods that are useful for understanding the structure and bonding in Zintl ions and related bare post-transition element clusters in approximate historical order. After briefly discussing the Zintl-Klemm model the Wade-Mingos rules and related ideas are discussed. The chapter concludes with a discussion of the jellium model and special methods pertaining to bare metal clusters with interstitial atoms.

Keywords: Jellium model · Metal clusters · Wade-Mingos rules · Zintl ions

Contents

R.B. King
Department of Chemistry, University of Georgia, Athens, GA 30602, USA
e-mail: rbking@chem.uga.edu

1 Introduction

Anionic species of the post-transition elements date back to the work of Johannis [1, 2] in 1891, who found that reduction of elemental lead with sodium in liquid ammonia gives a green solution apparently containing some type of lead anion. Subsequent detailed studies on this lead anion by Smyth in 1917 [3] suggested a $Pb_{2.25}^{-}$ stoichiometry, or Pb_9^{4-} with an integral number of lead atoms. As a result of these studies, as well as extensive subsequent work by Zintl and coworkers in the 1930s [4–7], species such as Na_4Pb_9 appeared in the inorganic books in the 1950s. At that time, the idea that lead could form a compound of any type in an oxidation state other than +2 or +4, particularly a negative oxidation state, seemed incomprehensible. Because of the pioneering work of Zintl in this area, bare anionic clusters are frequently known as Zintl ions.

Modern work on these and related bare post-transition element clusters began in the 1960s after Corbett and coworkers found ways to obtain crystalline derivatives of these post-transition element clusters by the use of suitable counterions. Thus, crystalline derivatives of the cluster anions had cryptate or polyamine complexed alkali metals as countercations [8]. Similarly, crystalline derivatives of the cluster cations had counteractions, such as $AlCl_4^{-}$, derived from metal halide strong Lewis acids [9]. With crystalline derivatives of these clusters available, their structures could be determined definitively using X-ray diffraction methods.

Early efforts to rationalize the observed formulas, shapes, and chemical bonding of these clusters used the Zintl–Klemm concept [10, 11] and the Mooser–Pearson extended $(8 - N)$ rule [12], where N refers to the number of valence electrons of the more electronegative metal in the intermetallic. However, the inadequacy of these simple models soon became apparent. At that point, the analogy between the bare post-transition metal clusters and polyhedral boranes was recognized so that the Wade–Mingos rules [13–16], which were developed to understand structure and bonding in polyhedral boranes and related compounds, could be extended to bare post-transition element clusters having structures similar to polyhedral boranes. The Wade–Mingos rules also provided a basis of extending the concept of aromaticity from two-dimensional planar hydrocarbons, such as the prototypical benzene, to three-dimensional polyhedral clusters. In addition, methods based on a graph–theoretical derivation [17] of the Wade–Mingos rules can be used to demonstrate the analogy between the aromaticity in planar hydrocarbons, such as benzene, and deltahedral boranes. In the context of polyhedral borane or cluster chemistry, the deltahedra, which are polyhedra with only triangular faces, are fundamental building blocks.

An essential feature of the Wade–Mingos rules [13–16] is the partitioning of the valence electrons of the cluster atoms into skeletal electrons and external electrons. For polyhedral boranes, the external electrons are used to form bonds to external groups, such as hydrogen atoms in the prototypical borane dianions $B_nH_n^{2-}$. However, for bare post-transition element clusters, these external electrons formally correspond to nonbonding electron pairs. Subsequent experimental and theoretical developments in bare post-transition element cluster chemistry indicate that this distinction is very important. In this connection, the most dramatic example compares the isovalent icosahedral species $B_{12}H_{12}^{2-}$ and Si_{12}^{2-}. Thus, although $B_{12}H_{12}^{2-}$ salts are extremely thermally and oxidatively stable species, the analogous Si_{12}^{2-} (or Ge_{12}^{2-}) clusters have never been synthesized. Theoretical studies also showed that the lowest energy structures for the germanium clusters Ge_n^{2-} (n=8 [18], 11 [19], and 14 [20]) are different from those of the isovalent polyhedral boranes $B_nH_n^{2-}$.

These observations suggested that the Wade–Mingos rules [13–16], like the earlier Zintl–Klemm concept [10, 11], are also inadequate to account for many aspects of structure and bonding in Zintl ions and related bare post-transition element clusters. In this connection, King and Silaghi-Dumitrescu [21] observed that the jellium model of physicists [22, 23] could also be applied to the Zintl ions and related bare post-transition element clusters. This model assumes that all of the valence electrons of the cluster atoms participate in the skeletal bonding including those allocated formally to the "external" lone pairs for the Wade–Mingos rules.

Meanwhile, experimental work on bare post-transition metal clusters led to the discovery of such clusters with interstitial atoms in the center of the cluster cavity. Relatively early examples of such species include the centered 10-vertex clusters $M@In_{10}^{10-}$ (M = Ni, Pd, Pt) found in intermetallic structures. Spectacular recent examples are the clusters $M@Ge_{10}^{3-}$ (M = Fe [24], Co [25]), containing the transition metal at the center of a Ge_{10} pentagonal prism. These latter examples are very provocative since the external Ge_{10} pentagonal prism has no triangular faces at all but only rectangular and pentagonal faces! This is in stark contrast to the fundamental deltahedra of borane chemistry in which all of the faces are triangles. Furthermore, recent theoretical studies suggest that in some cases the preferred cluster polyhedron for enclosing a central metal atom is simply the polyhedron with the largest internal volume rather than a polyhedron meeting the electronic requirements of either the Wade–Mingos rules [13–16] or the jellium model [21]. For the 10-vertex systems, this polyhedron is the pentagonal prism with the minimum number of edges and hence no triangular faces.

This chapter reviews the methods that are useful for understanding the structure and bonding in Zintl ions and related bare post-transition element clusters in approximate historical order. Thus, after briefly mentioning the early Zintl–Klemm concept, we proceed to discuss the Wade–Mingos rules and related ideas. We then conclude with a discussion of the jellium model and special methods pertaining to bare metal clusters with interstitial atoms.

2 Early Efforts to Rationalize the Structures of the Zintl Ions: The Zintl–Klemm Concept

Early efforts to rationalize the observed formulas and chemical bonding of Zintl ions and related species used initially the Zintl–Klemm concept [10, 11] and subsequently the Mooser–Pearson [12] extended (8 − N) rule. In this rule, N refers to the number of valence electrons of the more electronegative metal (and thus the anionic metal) in the intermetallic.

The intermetallic phase [26] Na_2Tl illustrates a simple application of the Zintl–Klemm concept to a group 13 metal cluster. Complete electron transfer from Na to Tl leads to the $(Na^+)_2Tl^{2-}$ formulation. The Tl^{2-} dianion is isoelectronic with group 15 elements and thus should form similar tetrahedral structures with six two-center two-electron bonds along the edges of the tetrahedron. Indeed, the Tl^{2-} anions in the Na_2Tl phase form $Tl_4{}^{8-}$ tetrahedra, similar to the isoelectronic P_4 and As_4 units in white phosphorus and yellow arsenic.

For a general formulation of the Zintl–Klemm concept, consider an intermetallic A_mX_n phase, where A is the more electropositive element, typically an alkali or an alkaline earth metal. Both A and X, viewed as individual atoms, are assumed to follow the octet rule leading to transfer of electrons from A to X, i.e., $A \rightarrow A^{p+}$, $X \rightarrow X^{q-}$, so that $|mp| = |nq|$. The anionic unit X^{q-} arising from this electron transfer is considered to be a pseudoatom, which exhibits a structural chemistry closely related to that of the isoelectronic elements [11]. Since bonding also is possible in the cationic units, the numbers of electrons involved in A–A and X–X bonds of various types (e_{AA} and e_{XX}, respectively) as well as the number of electrons e^* not involved in localized bonds can be generated from the numbers of valence electrons on A and X, namely e_A and e_X, respectively, by the following equations of balance:

$$me_A + ne_X + k = 8n \tag{1a}$$

$$k = -\sum e_{AA} + \sum e_{XX} - \sum e^* \tag{1b}$$

This simple Zintl–Klemm concept has some obvious limitations as illustrated by differences in the two MTl intermetallics (M = Na [27] and Cs [28]). Thus complete electron transfer from M to Tl leads to M^+Tl^-, where the Tl^- anion with four valence electrons is isoelectronic with a neutral group 14 atom and four bonds are needed to attain the octet configuration. Hence, the Tl^- anion should form structures similar to allotropes of carbon or heavier group 14 elements. Indeed, NaTl has a "stuffed diamond" structure [27] with internal Na^+ ions and an anionic $(Tl^-)_\infty$ lattice similar to diamond. However, the Tl^- anions in [CsTl] form tetragonally compressed octahedra [28] unlike any structures of the allotropes of carbon or its heavier congeners.

The efforts to rationalize the formulas and structures of Zintl ions and related species predated extensive definitive structural information on anionic post-transition metal clusters obtained by Corbett and his group in the 1970s [8, 9]. After enough such structural information on the bare post-transition metal clusters became available, the resemblance of their polyhedra to the known polyhedral boranes became apparent. For this reason, the simple Zintl–Klemm concept has been largely superseded by newer, more advanced descriptions of chemical bonding in such clusters, initially those applied to the polyhedral boranes.

3 The Relationship of Zintl Ions to Polyhedral Boranes: The Wade–Mingos Rules and the Aromaticity of Clusters

3.1 Application of the Wade–Mingos Rules to Polyhedral Boranes: Nido *and* Arachno *Borane Structures*

The next development in the understanding of structure and bonding in the Zintl ions recognized their relationship to the polyhedral boranes and the isoelectronic carboranes. Then the Wade–Mingos rules [13–16], which were developed to understand the structure and bonding in polyhedral boranes, could be extended to isovalent Zintl ions and related post-transition element clusters.

The basic building blocks of the polyhedral boranes and isoelectronic carboranes are the so-called most spherical or so-called *closo* deltahedra (Fig. 1) [29, 30], in which the vertices are as nearly similar as possible. For such polyhedra having from 6 to 12 vertices, this means that all vertices have degrees of either 4 or 5 with only one exception, namely the 11-vertex deltahedron, where a deltahedron with only degree 4 and 5 vertices is topologically impossible [31]. Therefore, the most spherical 11-vertex deltahedron has one degree 6 vertex in addition to two degree 4 vertices and eight degree 5 vertices (Fig. 1). The conclusion from the Wade–Mingos rules [13–16] is that deltahedral boranes with n vertices are particularly stable if they have $2n + 2$ skeletal electrons. Thus the dianions $B_nH_n^{2-}$, the carborane monoanions $CB_{n-1}H_n^-$, the neutral carboranes $C_2B_{n-2}H_n$, and their substitution products are particularly stable.

To apply the Wade–Mingos rules [13–16], consider a *closo* borane or isoelectronic carborane deltahedron with n vertices in which all of the vertices are boron or carbon atoms having four-orbital sp^3 valence manifolds. Using one of the four valence orbitals on each vertex atom for external bonding leaves three internal orbitals per vertex atom for a total of $3n$ orbitals available for the internal skeletal bonding of the deltahedron [17]. The BH and CH vertices in polyhedral borane and carborane derivatives are seen to be donors of two and three skeletal electrons, respectively, after allowing for the single valence electron of each vertex atom required for the two-center two-electron bond to the external hydrogen atom.

The Wade–Mingos model [13–16] for the skeletal bonding in deltahedral boranes with n vertices is based on the partitioning of the $3n$ internal orbitals of the vertex atoms into two twin internal or tangential orbitals per vertex atom and a single unique internal or radial orbital per vertex atom leading to a total of $2n$ tangential orbitals and n radial orbitals. Pairwise overlap between the $2n$ tangential orbitals is responsible for the surface bonding of the deltahedron and splits these $2n$ tangential orbitals into n bonding and n antibonding orbitals corresponding to n two-center two-electron (2c–2e) bonds along a Hamiltonian circuit in the surface of the deltahedron [17]. In this connection, a Hamiltonian circuit in a polyhedron starts at a given vertex and visits all of the other vertices without visiting any vertex twice. This surface bonding model is a localized model; the actual surface bonding is a

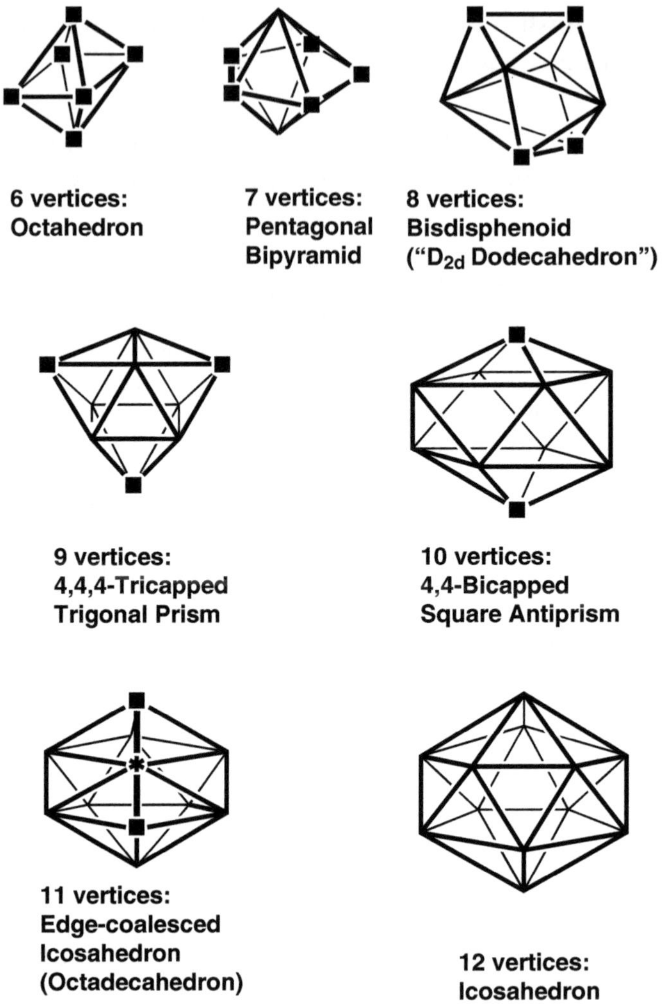

Fig. 1 The "most spherical" (*closo*) polyhedra found in borane derivatives such as $B_nH_n^{2-}$. In Figs. 1 and 2, the degree 6, 4, and 3 vertices are indicated by an *asterisk*, a *square*, and a *triangle*, respectively, and degree 5 vertices are *unmarked*

resonance hybrid using all of the possible Hamiltonian circuits on the *closo* deltahedron.

The n radial orbitals in a deltahedral borane with n vertices are oriented toward the center of the deltahedron. Mutual overlap of these radial atomic orbitals leads to an n-center core bond generating one additional bonding molecular orbital and $n - 1$ additional antibonding molecular orbitals. This n-center core bond supplements the surface bonding [17]. The combination of the surface and core bonding thus leads to a total of $n + 1$ bonding orbitals. Filling each of these bonding orbitals with an electron pair leads to the $2n + 2$ skeletal electrons required by the Wade–Mingos rules [13–16] for a globally delocalized three-dimensional aromatic *closo* deltahedron.

Now let us apply similar reasoning to *nido* boranes of the general formula B_nH_{n+4}. Such boranes have structures based on a polyhedral fragment obtained by removal of one vertex and its associated edges from a closed deltahedron with $n + 1$ vertices to generate an "open face" or "hole." The *nido* boranes in Fig. 2 are all stable compounds. The B_5H_9 structure is based on a B_5 square pyramid obtained from a B_6 octahedron by removing one of the vertices, thereby generating the square base of the pyramid. Similarly, the B_6H_{10} structure is based on a B_6 pentagonal pyramid obtained from a B_7 pentagonal bipyramid (Fig. 1) by removing a degree 5 vertex, thereby generating the pentagonal base of the pyramid. The B_8H_{12} structure is generated from a 9-vertex deltahedron with a single degree 6 vertex by removing that degree 6 vertex. Note that in this case the original 9-vertex deltahedron, with a single degree 6 vertex, is not the most spherical 9-vertex deltahedron, namely the tricapped trigonal prism with only degree 4 and 5 vertices (Fig. 1). The $B_{10}H_{14}$ structure is generated from the most spherical 11-vertex deltahedron (Fig. 1) by removing the unique degree 6 vertex. Now assume that each boron atom has an external terminal hydrogen atom thereby accounting for all except four hydrogen atoms in the B_nH_{n+4} structure. These four "extra" hydrogen atoms generally appear as bridges across pairs of boron atoms bordering the open face. These *nido* polyhedra with n vertices have the same number of bonding molecular orbitals as the $(n + 1)$-vertex deltahedra from which they are derived by single vertex removal. Thus, they have $(n + 1) + 1$ bonding orbitals leading to systems with $2n + 4$ skeletal electrons.

Removal of two vertices from a deltahedron leads to *arachno* polyhedra having $2n + 6$ skeletal electrons according to the Wade–Mingos rules [13–16]. Among the many theoretical possibilities for *arachno* polyhedra, the pentagonal antiprism, generated by removal of a pair of antipodal vertices from an icosahedron, is the most significant in bare post-transition metal cluster chemistry.

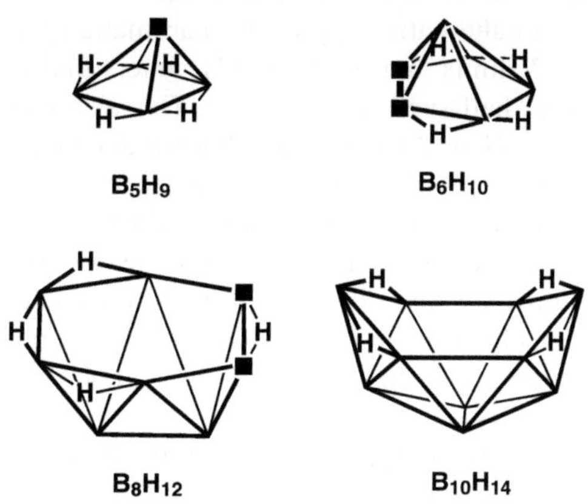

Fig. 2 The structures of the known *nido* boranes B_nH_{n+4} (n = 5, 6, 8, 10). In each structure, the four bridging hydrogen atoms are shown but not the terminal hydrogen atoms

3.2 Aromaticity in Boranes and Related Clusters

The *closo* boranes $B_nH_n^{2-}$ ($6 \leq n \leq 12$), along with their isoelectronic counterparts, the carboranes, $CB_{n-1}H_n^-$, and $C_2B_{n-2}H_n$, exemplify three-dimensional aromaticity [17, 32]. The bonding topology in such systems is analogous to that in familiar two-dimensional planar aromatic hydrocarbons such as benzene [17]. Thus, the pair of tangential orbitals from each boron vertex in the deltahedral boranes forming the surface bonding is analogous to the pair of sp^2 hybrid orbitals from a carbon vertex of benzene forming σ-bonds with the two adjacent carbon atoms in the six-membered ring. Similarly, the single radial orbital from each boron vertex in the deltahedral boranes forming the multicenter two-electron core bond in the center of the deltahedron is analogous to the single p orbital on each carbon atom of the benzene ring overlapping with hexagonal topology to form the 6 π-electron system in benzene. The number of electrons in the π-systems of the planar aromatic hydrocarbons follows the famous Hückel $4k + 2$ rule where $k = 0$ for the cyclopropenyl cation ($C_3H_3^+$); $k = 1$ for the cyclopentadienide anion ($C_5H_5^-$), benzene (C_6H_6), and the tropylium cation ($C_7H_7^+$); and $k = 2$ for the cyclooctatetraene dianion ($C_8H_8^{2-}$). All of these planar hydrocarbons are species stabilized by aromaticity. For the three-dimensional deltahedral boranes, the counterpart of the π-system of planar hydrocarbons is the n-center core bond, which always contains two electrons. Thus, if the Hückel $4k + 2$ rule is applied to deltahedral boranes, the value of k is seen to be zero for all known borane dianions $B_nH_n^{2-}$ and the isoelectronic carboranes $C_2B_{n-2}H_n$ ($6 \leq n \leq 14$).

More advanced mathematical aspects of aromaticity are given in other references [33, 34]. Some alternative methods beyond the scope of this chapter for the study of aromaticity in deltahedral molecules include tensor surface harmonic theory [35–38] and the related Hirsch $2(N + 1)^2$ electron-counting rule for spherical aromaticity [39]. The topological solitons of nonlinear field theory related to the Skyrmions of nuclear physics have also been used to describe aromatic cluster molecules [40].

An alternative approach to aromaticity in some deltahedral boranes is based on the Jemmis and Schleyer [41] interstitial electron rule, originally introduced for *nido* "half-sandwich" species. This also relates aromaticity in deltahedral boranes to the $4k + 2$ Hückel rule, but for deltahedral boranes k is typically 0.

To apply the Jemmis–Schleyer interstitial electron rule [41], the *closo* $B_nH_n^{2-}$ dianions (their isoelectronic analogs are treated similarly) are dissected conceptually into two BH^- "caps" and one or two constituent $(BH)_n$ rings. The BH^- caps contribute three interstitial electrons each, but the rings (which, formally, have zero electrons in the π MOs) contribute none. Hence, six electrons, described as "interstitial", link the bonding symmetry-adapted cap and ring orbitals together perfectly.

The bonding analysis of the 50 $B_{12}H_{12}^{2-}$ valence electrons is illustrative. After the conceptual dissection into two BH caps and two $(BH)_5$ rings, two electrons each are assigned to the 12 BH bonds and to the 10 BB ring bonds. This leaves six electrons ($4k + 2$ for $k = 1$) for the interstitial bonding, which holds the rings and caps together. Icosahedral symmetrization then completes the description.

The Jemmis–Schleyer interstitial electron rules [41] are directly applicable to 5-, 6-, and 7-vertex deltahedra (which have one ring), and to 10-, 11-, and 12-vertex deltahedra (which have two rings) but are less obvious for 8- and 9-vertex deltahedra. However, in $B_8H_8^{2-}$ two $B_2H_2^-$ caps supply three interstitial electrons each to the central D_{2d}-puckered B_4H_4 ring of $B_8H_8^{2-}$. In the D_{3h} $B_9H_9^{2-}$, the six interstitial electrons are provided by the three, widely separated central BH "caps". In the latter, the electrons from the –2 charge on the ion can be assigned to a weak three-center two-electron bond involving the three BH "caps".

Ring current theory applied to chemical shifts can be used widely as an aromaticity criterion. Thus, proton NMR signals (and those of other elements) located in the "shielding cone" above or even inside an aromatic ring are shifted upfield, whereas nuclei outside the ring plane are "deshielded" to positions downfield. This is the basis of the nucleus independent chemical shift (NICS) criterion of aromaticity, introduced by Schleyer and collaborators in 1996 [42]. In this connection, many electronic structure programs, widely used to compute chemical shifts of atoms [43], can be used routinely to compute NICS using "ghost atoms" at chosen points. The signs of absolute shieldings obtained in this manner are merely reversed to conform to the chemists' signs convention (negative upfield, positive downfield). Since NICS, like chemical shifts, are influenced by the local magnetic effects of two-center two-electron bonds, it is useful interpretively to "dissect" the total NICS values into contributions of localized orbitals [44] (which are like chemical bonds, lone pairs, and core electrons) and contributions of the individual canonical MOs [45, 46]. In addition to giving the π contributions of an arene, the other MOs also provide instructive and often surprising insights [47]. These LMO and MO analyses have been used extensively to provide complimentary interpretive information.

The problem of distinguishing "aromaticity" in species merely having three-dimensional geometries from situations in which the "aromaticity" requires or at least strongly depends on the special nature or characteristics of three-dimensional systems can be addressed easily by NICS evaluations. The circulation of "mobile electrons" in a more conventional aromatic system involves a single circuit (either in a plane or traversing an undulating geometry). The "mobile electrons" in a strictly defined three-dimensional aromatic system use three circuits (representing the principal dimensions) for their delocalization. In this connection, Schleyer, Najafian et al. [48–50] have used computed geometric, energetic, and magnetic properties to quantify the aromaticity of the *closo* boranes $B_nH_n^{2-}$ ($6 \leq n \leq 12$), and their isoelectronic counterparts, namely the $CB_{n-1}H_n^-$ and $C_2B_{n-2}H_n$ carboranes and the $NB_{n-1}H_n$ azaboranes [51].

3.3 Application of the Wade–Mingos Rules to Bare Post-Transition Metal Clusters

Information on the numbers of skeletal electrons contributed by each cluster vertex is necessary to apply the Wade–Mingos rules [13–16] to bare clusters of the post-transition elements. The rules discussed above for polyhedral boranes can be adapted to bare post-transition metal vertices as follows:

1. The post-transition metals use a four-orbital sp^3 valence orbital manifold. The inner shell d orbitals are assumed not to be involved in the bonding but instead comprise nonbonding electron pairs.
2. Three orbitals of each bare metal vertex atom are required for the internal orbitals (two twin internal orbitals and one unique internal orbital). This leaves one external orbital for a nonbonding lone pair.

Application of this procedure to post-transition metal clusters indicates that bare Ga, In, and Tl vertices contribute one skeletal electron; bare Ge, Sn, and Pb vertices contribute two skeletal electrons; bare As, Sb, and Bi vertices contribute three skeletal electrons; and bare Se and Te vertices contribute four skeletal electrons in two- and three-dimensional aromatic systems. Thus, Ge, Sn, and Pb vertices are isoelectronic with BH vertices and As, Sb, and Bi vertices are isoelectronic with CH vertices.

Some examples of bare ionic post-transition metal clusters are depicted in Fig. 3. Their chemical bonding topologies can be analyzed as follows:

1. *Square*. Bi_4^{2-}, Se_4^{2+}, and Te_4^{2+} are isoelectronic and isolobal with the delocalized planar cyclobutadiene dianion. They have 14 skeletal electrons (e.g., for Bi_4^{2-}: (4) (3) + 2 = 14) corresponding to eight electrons for the four σ-bonds and six electrons for the π-bonding.
2. *Butterfly*. While $Tl_2Te_2^{2-}$ has a (2) (1) + (2) (4) + 2 = 12 skeletal electron count isoelectronic and isolobal with *neutral* cyclobutadiene, it undergoes a different Jahn–Teller-like distortion to the butterfly structure found experimentally by Burns and Corbett [52].
3. *Tetrahedron*. $Sn_2Bi_2^{2-}$ and $Pb_2Sb_2^{2-}$ have (2) (2) + (2) (3) + 2 = 12 skeletal electrons for tetrahedra analogous to P_4 and organic C_4R_4 tetrahedrane.
4. *Trigonal Bipyramid*. Sn_5^{2-}, Pb_5^{2-}, and Bi_5^{3+} have 12 skeletal electrons [e.g., (5) (2) + 2 = 12 for Sn_5^{2-} and Pb_5^{2-}] and are analogous to the trigonal bipyramidal $C_2B_3H_5$ carborane.
5. *Seven-Vertex Structures*. As_7^{3-} and Sb_7^{3-} have the C_{3v} structure depicted in Fig. 2 and the correct (4) (3) + (3) (1) + 3 = 18 skeletal electron count for bonds along the nine edges derived by considering the three vertices of degree 2 to use two internal orbitals each and the four vertices of degree 3 to use three internal orbitals each.
6. *Capped Square Antiprism*. Ge_9^{4-}, Sn_9^{4-}, and Pb_9^{4-} have the (9) (2) + 4 = 22 = $2n + 4$ skeletal electrons required for an $n = 9$ vertex C_{4v} nido polyhedron having 12 triangular faces and one square face.

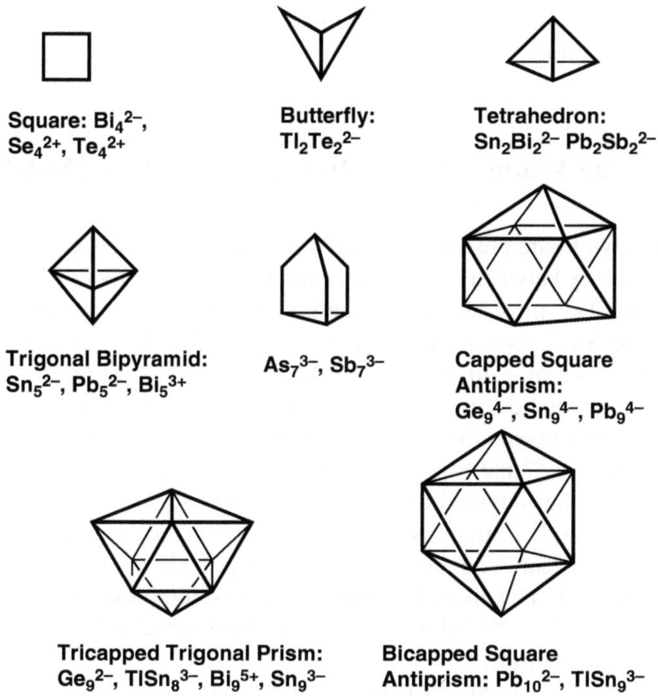

Fig. 3 Examples of the shapes of some Zintl ions and related clusters

7. *Tricapped Trigonal Prism.* Ge_9^{2-} and $TlSb_8^{3-}$ have the $2n + 2 = 20$ skeletal electrons required for an $n = 9$ vertex globally delocalized D_{3h} deltahedron analogous to $B_9H_9^{2-}$ [53]. The bismuth cluster Bi_9^{5+} anomalously has (9) (3) − 5 = 22 rather than the expected 20 skeletal electrons in most derivatives suggesting [54] incomplete overlap of the unique internal orbitals directed toward the core of the deltahedron. Only in the unusual environment of the recently synthesized [55] $Bi_{18}Sn_7Br_{24}$ (= $(Bi_9^{5+})_2[Sn_7Br_{24}^{10-}]$) is the Bi_9^{5+} cluster cation the C_{4v} *nido* capped square pyramid predicted by the Wade–Mingos rules [13–16]. The paramagnetic anion Sn_9^{3-} has (9) (2) + 3 = 21 skeletal electrons including one extra electron for a low-lying antibonding orbital analogous to radical anions formed by stable aromatic hydrocarbons such as naphthalene and anthracene.
8. *Bicapped Square Antiprism.* Pb_{10}^{2-} has the (10) (2) + 2 = 22 = $2n + 2$ skeletal electrons required for an $n = 10$ vertex globally delocalized D_{4d} deltahedron [56] (cf. the bicapped square antiprism in Fig. 3) analogous to that found in the $B_{10}H_{10}^{2-}$ anion [57]. $TlSn_9^{3-}$ is isoelectronic with Pb_{10}^{2-} and also has a bicapped square antiprism structure.

The Wade–Mingos rules appear to be applicable to the above clusters when they are polyhedra in which each vertex has degree of at least 4 (e.g., the 9-vertex capped square antiprism and tricapped trigonal prism clusters and the 10-vertex bicapped square antiprism clusters).

4 Beyond the Wade–Mingos Rules: The Jellium Model

As the chemistry of bare post-transition element clusters developed further, limitations in this analogy between such clusters and polyhedral boranes became increasingly apparent. The most dramatic example occurs in the regular icosahedral systems, where Si_{12}^{2-} isovalent to the very stable $B_{12}H_{12}^{2-}$ remains unknown. The reason for this major difference was discovered relatively recently [58] when $B_{12}H_{12}^{2-}$ was found to be aromatic but Si_{12}^{2-} to be antiaromatic using the NICS test for aromaticity discussed above [42]. A group-theory basis for this difference in behavior of $B_{12}H_{12}^{2-}$ and Si_{12}^{2-} was subsequently developed [59]. Furthermore, the NICS test for the icosahedral Group 14 clusters suggests that Ge_{12}^{2-} is less antiaromatic than Si_{12}^{2-}, Sn_{12}^{2-} is essentially nonaromatic, and Pb_{12}^{2-} is aromatic [60]. This is consistent with the detection of E_{12}^{2-} (E = Sn, Pb) in gas-phase experiments [61, 62], as well as the isolation of crystalline salts of the icosahedral anions $M@Pb_{12}^{2-}$ (M = Ni, Pd, Pt), which have been structurally characterized by X-ray crystallography [63].

The group 13 metals, particularly Ga, In, and Tl, are another source of interesting bare metal cluster structures differing considerably from any of the polyhedral borane structures (Fig. 4) [64]. If the Wade–Mingos rules [13–16] for counting skeletal electrons are used, bare group 13 metal vertices then have only one electron available for skeletal bonding after using two of their three valence electrons for the external lone pair. Because of the resulting need for extra skeletal electrons, the bare group 13 metal clusters are generally highly charged ionic species. They frequently occur in intermetallics with alkali metals as the source of the extra electrons. However, clusters constructed exclusively from bare group 13 metals, even if they bear relatively high negative charges, are still likely to be highly hypoelectronic relative to the $2n + 2$ skeletal electrons required by the Wade–Mingos rules for a borane-like most spherical deltahedron. The hypoelectronicity of many of the polyhedra found in such group 13 metal clusters appears to be relieved by pushing one or more vertices toward the center of the polyhedron [indicated by spades (♠) in Fig. 4] [65]. The "flattening" process pushes the otherwise external lone pair of the flattened vertex inside the polyhedron so that the originally "external" lone pair now becomes an "internal" lone pair to participate in the skeletal bonding.

One of the most interesting bare Group 13 metal clusters is the first one to be discovered [66], namely the 11-atom cluster In_{11}^{7-} in the intermetallic K_8In_{11}. Analogous 11-vertex clusters were subsequently synthesized containing gallium [67] and thallium [68]. The In_{11}^{7-} cluster has 11 + 7 = 18 skeletal electrons = 2 $n - 4$ for $n = 11$. It is thus a highly hypoelectronic system relative to the $2n + 2$ skeletal electron deltahedral boranes $B_nH_n^{2-}$. The polyhedron found in In_{11}^{7-} is

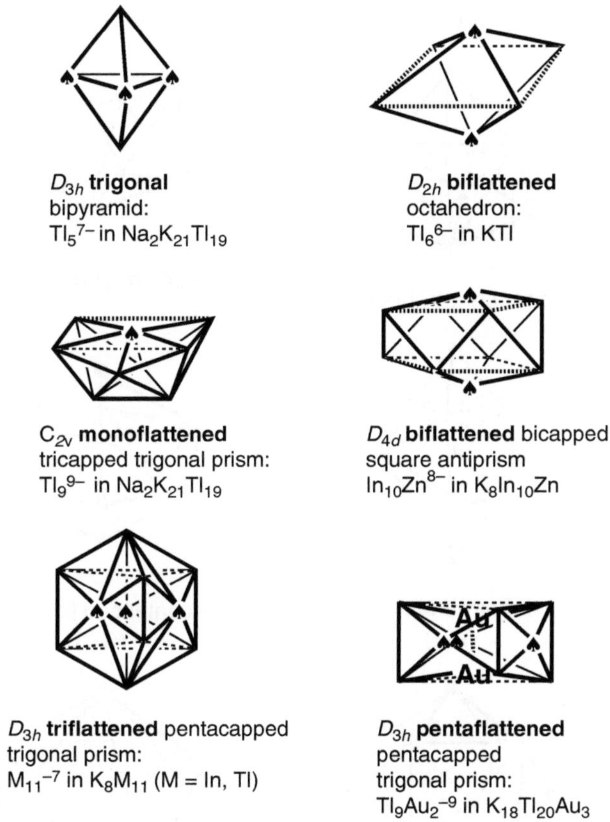

Fig. 4 Unusual polyhedra found in Group 13 intermetallics. The flattened vertices in these polyhedra are indicated by *spades*

derived from a D_{3h} pentacapped trigonal prism by elongating the horizontal (triangular face) edges of the underlying trigonal prism (Fig. 4). This polyhedron is very different from the C_{2v} polyhedron found in $B_{11}H_{11}^{2-}$ (Fig. 1). This unusual In_{11}^{7-} polyhedron is obviously a very favorable polyhedron for highly reduced group 13 cluster systems since K_8In_{11} is made from the very unselective reaction of fusing a 40/60 potassium–indium mixture at 530°C [66].

To understand the unusual structures found in bare post-transition element clusters and underlying reasons for differences in the structures between analogous boranes and post-transition element clusters, we carried out extensive density functional theory (DFT) studies of bare germanium clusters during the past decade [18–20, 69–72]. In three cases (Fig. 5), these studies predict lowest energy structures for Ge_n^{2-} clusters that are different from the well-established structures of the isovalent boranes $B_nH_n^{2-}$ and carboranes $C_2B_{n-2}H_n$. Thus, the lowest energy structure for Ge_8^{2-} is a T_d tetracapped tetrahedron, whereas that for $B_8H_8^{2-}$ is a D_{2d}

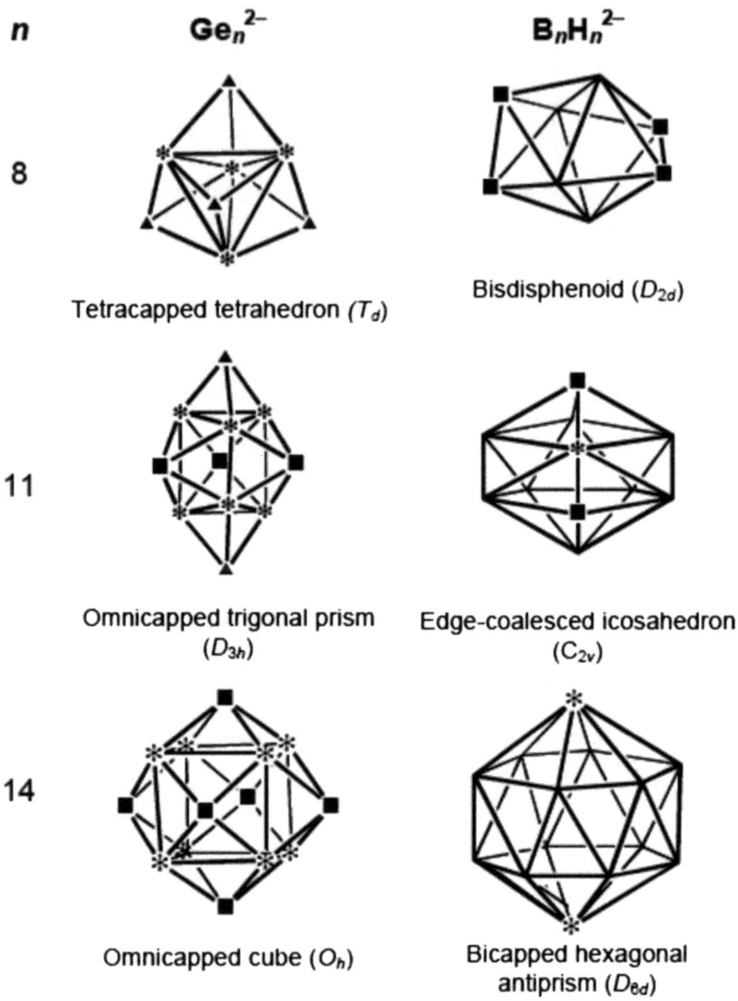

Fig. 5 Examples of different lowest energy deltahedra for isoelectronic Ge_n^{2-} and $B_nH_n^{2-}$ derivatives (n = 8, 11, 14)

bisdisphenoid. The D_{2d} bisdisphenoid structure for Ge_8^{2-} lies ~4 kcal/mol above the T_d tetracapped tetrahedron global minimum [18]. Similarly, the lowest energy structure for Ge_{11}^{2-} is a D_{3h} pentacapped trigonal prism, whereas that for $B_{11}H_{11}^{2-}$ is a C_{2v} deltahedron. The C_{2v} deltahedral structure for Ge_{11}^{2-} lies ~5 kcal/mol above the D_{3h} pentacapped trigonal prism global minimum [19]. For Ge_{14}^{2-}, the structure greatly preferred energetically is a very symmetrical O_h omnicapped cube, whereas that predicted for $B_{14}H_{14}^{2-}$ and observed experimentally [73] in a carborane of the type $R_2C_2B_{12}H_{12}$ is a D_{6d} bicapped hexagonal antiprism. Note that in these three cases, the favored borane deltahedra (Figs. 1 and 5) maximize

Fig. 6 Comparison of the electronic energy levels in an atom and a jellium sphere

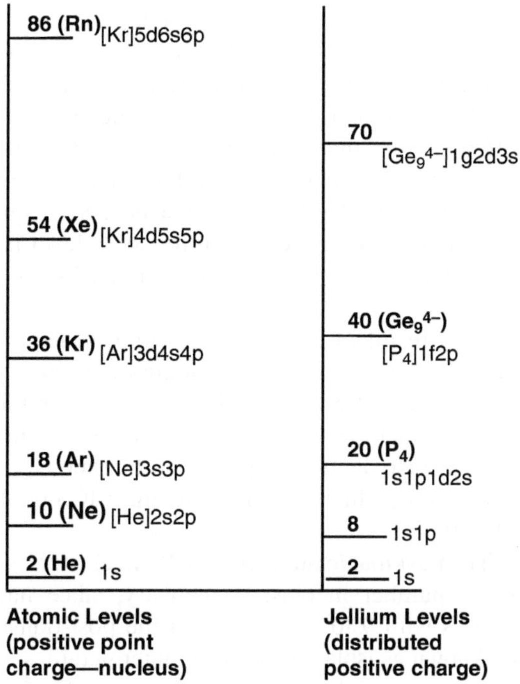

the numbers of degree 5 vertices, whereas the favored bare germanium cluster deltahedra have degree 4 and 6 vertices rather than degree 5 vertices.

These examples indicate significant limitations in the Wade–Mingos rules [13–16] to predict structures of bare post-transition metal clusters. We therefore adapted the "magic numbers" of the jellium model used by physicists for gas-phase metal clusters [22, 23] to understand structure and bonding in the bare post-transition metal clusters [21]. In this connection, the jellium model arises from the recognition of certain bare spherical free-electron metal clusters as "superatoms" among which the 40 valence electron species Al_{13}^-, observed in gas-phase experiments [74], is the prototypical example. However, although in an atom the positive charge of a nucleus is concentrated at a central point, in a cluster the positive charge can be assumed to be distributed over a positive ion core of the size of the cluster leading to a so-called jellium sphere. Both cases lead to discrete electronic energy levels. However, the different distributions of the positive charges lead to different ordering of the electronic energy levels (Fig. 6).

The filling of atomic orbitals follows an $(n + \ell, n)$ orbital scheme known as the Madelung [75–77] or Klechkovskii [78] rule. In this orbital scheme, the electron occupies free states with the smallest value of the sum $N = n + \ell$ of the principal quantum number n and the azimuthal quantum number ℓ according to the Pauli exclusion principle. In the presence of several states with identical N, the state with the smallest n (and hence with the largest ℓ) is filled first. This leads to closed shell configurations with 2, 10, 18, 36, 54, and 86 electrons corresponding to the noble gases, which are the least chemically reactive of all of the elements (Fig. 6).

The order of filling of the free electronic states in a jellium sphere is quite different from that of a free atom (Fig. 6). Although in a free atom the 2*s* level immediately follows the 1*s* level, in a jellium sphere the 1*p* and 1*d* levels follow the 1*s* level before the 2*s* level is reached. Subsequently in the free atom only the 2*p* level follows the 2*s* level before the 3*s* level is reached. However, in the jellium model the 2*s* level is followed by the 1*f*, 2*p*, 1*g*, and 2*d* levels before the 3*s* level is reached.

As a result of this electron-filling scheme for jellium clusters, the "magic" numbers for closed shell configurations in a jellium cluster are very different from those in free atoms. The first magic number of chemical significance in a jellium sphere is the 20 valence electron configuration of white phosphorus P_4 and other isoelectronic species of the type E_4 (E = As, Sb, Bi) and E_4^{4-} (E = Si, Ge, Sn, Pb), which are shown by the NICS method to be highly aromatic systems [39, 79].

The next magic number for jellium clusters is 40. This is a particularly important magic number in cluster chemistry, since numerous 40 valence electron bare clusters with 9 to 11 vertices of the post-transition elements in Groups 13 to 15 are known as isolable species in intermetallics or "salts" with suitable counterions. Examples of such species include In_{11}^{7-}, Ge_9^{4-}, and Bi_9^{5+}, all of which have been isolated in intermetallics (for In_{11}^{7-}) or as stable salts with suitable counterions (Ge_9^{4-} and Bi_9^{5+}) and characterized by X-ray crystallography.

Comparison between the jellium model [21] and the Wade–Mingos rules [13–16] is instructive. The Wade–Mingos rules suggest spherical deltahedral structures (Fig. 1) for clusters having n vertices and $2n + 2$ skeletal electrons. Such a skeletal electron count excludes "external" electron pairs or electrons involved in bonding to external atoms or groups such as the hydrogen atoms in the deltahedral borane anions $B_nH_n^{2-}$. However, the jellium model counts all of the valence electrons of the cluster atoms including those that would be considered as "external" lone pairs by the Wade–Mingos rules. For example, the capped square antiprismatic Zintl ion [80] Ge_9^{4-} is counted as a 22 skeletal electron *nido* system with one "open" nontriangular face by the Wade–Mingos rules ($22 = 2n + 4$ for $n = 9$). However, Ge_9^{4-} is counted as a 40 valence electron system in the jellium model and thus has a favorable jellium closed shell configuration. The difference in Ge_9^{4-} between the 40 valence electron count of the jellium model and the 22 skeletal electron count used for the Wade–Mingos rules consists of the 18 electrons from the nine external lone pairs, one electron pair for each germanium vertex. Note that the jellium model is only directly relevant when these external electrons are available for the cluster bonding and therefore only applies to bare post-transition element clusters without external groups. Thus, the jellium model does not supplant the Wade–Mingos rules in borane chemistry, since the vertex boron atoms in polyhedral borane structures are always bonded to hydrogen or other external groups and thus do not have external lone pairs.

5 Clusters with Interstitial Atoms: Available Internal Volume Can Override Electronic Considerations

Recently some Zintl cluster ions have been shown to accommodate interstitial transition metal and post-transition metal atoms [81–83]. The smallest Zintl ion polyhedron shown to accommodate an interstitial transition metal ion is the 9-vertex tricapped trigonal prism (Fig. 1) in the derivatives $Cu@E_9^{3-}$ (E = Sn, Pb) [84]. These clusters are isoelectronic with neutral E_9^{4-} since the interstitial copper atom donates one skeletal electron to become effectively Cu^+ with a filled d^{10} shell. Larger deltahedra (Fig. 1) that can accommodate an interstitial atom include the bicapped square antiprism in $Ni@Pb_{10}^{2-}$ [85] and the icosahedron in $M@Pb_{12}^{2-}$ (M = Ni, Pd, Pt) [63, 86] and $Ir@Sn_{12}^{3-}$ [87]. In $Ir@Sn_{12}^{3-}$, the interstitial iridium atom is formally an anion Ir^- with a filled d^{10} shell. A pair of interstitial nonbonded palladium atoms (Pd $\cdots$ Pd = 3.384 Å) inside a prolate ellipsoidal 18-vertex deltahedron is found in the clusters $(Pd)_2@E_{18}^{4-}$ (E = Ge [88], Sn [89]). This structure is very different from the $Sn[Ni@Sn_8]_2^{4-}$ structure [90] of $Ni_2Sn_{17}^{4-}$. In the latter structure, two $Ni@Sn_9$ polyhedra derived from the tricapped trigonal prism share a tin vertex corresponding to one of the caps of the underlying trigonal prism. This central tin atom is eight-coordinate with square antiprismatic coordination.

The centered 10-vertex polyhedra are of particular interest since the shapes of the outer 10-vertex polyhedron depends on the interstitial atom and the electron count. In fact, four very different 10-vertex polyhedra (Fig. 7) have all been shown to form stable isolable species containing interstitial transition or post-transition metal atoms. These polyhedra include structures with three-, four-, or fivefold symmetry. Thus for the ions $M@In_{10}^{10-}$ (M = Ni, Pd, Pt) found in the intermetallics $K_{10}In_{10}M$, the In_{10} polyhedron is a C_{3v} tetracapped trigonal prism [91].

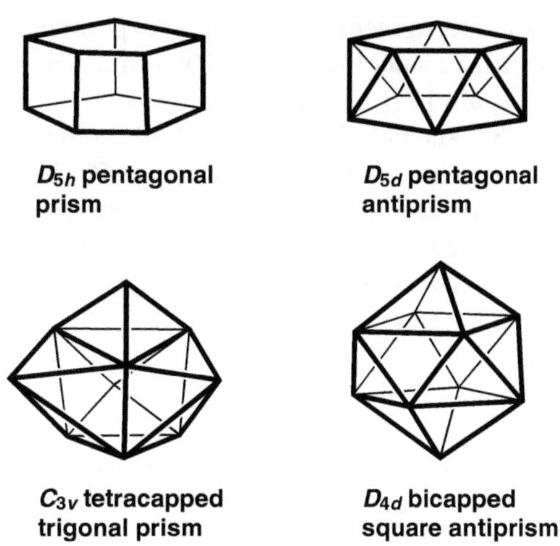

Fig. 7 The four 10-vertex polyhedra found in stable interstitial $M@Ge_{10}^{z}$ clusters

However, for the isoelectronic ion $Zn@In_{10}^{8-}$ found in the intermetallic $K_8In_{10}Zn$, the In_{10} polyhedron is a D_{4d} bicapped square antiprism [92]. Furthermore, the Bi_{10} polyhedron in the 26 skeletal electron $Pd@Bi_{10}^{4+}$ found in $Bi_{14}PdBr_{16}$ (= $[Pd@Bi_{10}][BiBr_4]_4$) [93] is different from either of these polyhedra but instead is a D_{5d} pentagonal antiprism. Finally, the spectacular clusters $M@Ge_{10}^{3-}$ (M = Fe [24], Co [25]) were recently synthesized in which the transition metal is in the center of a pentagonal prism. The last example is particularly interesting since the pentagonal prism has no triangular faces at all but only rectangular and pentagonal faces. This is in complete contrast to the borane deltahedra (Fig. 1) as well as other post-transition metal cluster deltahedra (Fig. 3) in which all of the faces are triangles.

To elucidate the factors determining the preferred shapes of metal-centered clusters, we carried out numerous DFT calculations on various $M@E_{10}^{z}$ derivatives [94–96]. The first such study [94] reproduced experimental results indicating different preferred deltahedra for the isoelectronic $Ni@Ge_{10}$ and $Zn@Ge_{10}^{2+}$, which were used as more computationally feasible models for $Ni@In_{10}^{10-}$ and $Zn@In_{10}^{8-}$ found in the intermetallics $K_{10}In_{10}Ni$ [91] and $K_8In_{10}Zn$ [92]. The germanium clusters were used as isoelectronic models for the indium clusters because of computational difficulties associated with the high negative charges on the indium clusters. In counting skeletal electrons in these systems, the interstitial nickel atom, with no electrons in excess of its filled d^{10} shell, is assumed to be a zero skeletal electron donor to the cluster system [97]. However, the interstitial zinc atom is considered to be Zn^{2+}, the normal oxidation state of zinc, so that an interstitial zinc atom is a two skeletal electron donor to the cluster structure. For the nickel-centered ion $Ni@Ge_{10}^{10-}$, the C_{3v} global minimum tetracapped trigonal prism structure is predicted to lie 4.6 kcal/mol in energy below the corresponding D_{4d} bicapped square antiprismatic structure (Fig. 7). However, for the zinc-centered ion $Zn@Ge_{10}^{8-}$, the relative order of the two structure types is reversed with the global minimum D_{4d} bicapped square antiprismatic structure lying 1.0 kcal/mol below the C_{3v} tetracapped trigonal prism structure. This is in complete agreement with the experimentally observed structures of the isoelectronic $Ni@In_{10}^{10-}$ and $Zn@In_{10}^{8-}$ in the intermetallics $K_{10}In_{10}Ni$ [91] and $K_8In_{10}Zn$ [92], respectively.

The second DFT study [95] compared nickel, palladium, and platinum as interstitial atoms in endohedral $M@Ge_{10}^{z}$ clusters (M = Ni, Pd, Pt; $z = 0, -2, -4$). The most interesting finding from this work was the observation of pentagonal prismatic structures as the global minima for $M@Ge_{10}^{z}$ (M = Pd, Pt, $z = -2, -4$). The final version of this theoretical manuscript was submitted for publication some months before appearance of the publications on the experimental work on the synthesis and structural characterization of the pentagonal prismatic clusters $M@Ge_{10}^{3-}$ (M = Fe [24], Co [25]).

The third DFT study [96] was motivated by the existence of the stable pentagonal antiprismatic cluster [93] $Pd@Bi_{10}^{4+}$ found in $Bi_{14}PdBr_{16}$. This cluster is an example of a cluster with an interstitial atom that obeys the Wade–Mingos rules [13–16]. Thus, the pentagonal antiprism is derived from an icosahedron by removal of a pair of antipodal vertices. The Wade–Mingos rules consider the pentagonal antiprism to be a 10-vertex *arachno* system expected to have $2n + 6$ skeletal electrons (= 26 electrons for $n = 10$). Since the bismuth atoms each contribute three skeletal electrons and the palladium atom contributes zero skeletal electrons because of its filled d^{10} shell [97], the cation $Pd@Bi_{10}{}^{4+}$ becomes a 26 skeletal electron system after considering its +4 charge. The DFT study [96] considered all nine possible combinations of the centered pnictogen clusters $M@Pn_{10}{}^{4+}$ (M = Ni, Pd, Pt; Pn = As, Sb, Bi). Among these nine combinations, only $Pd@Bi_{10}{}^{4+}$ and $Pt@Bi_{10}{}^{4+}$ were found to have the pentagonal antiprism as the lowest energy structure. The lowest energy structures for $M@Sb_{10}{}^{4+}$ (M = Ni, Pd, Pt) and $Ni@As_{10}{}^{4+}$ are derived from a C_{3v} tetracapped trigonal prism. A relatively large palladium or platinum atom does not fit very well into the smallest of the Pn_{10} clusters, namely the As_{10} cluster, as indicated by fragmentation of the original As_{10} polyhedron into two smaller pieces in the lowest energy structures.

A general conclusion from these DFT studies on $M@E_{10}$ clusters of various types is that the Wade–Mingos rules [13–16] are most applicable when the E_{10} unit has as large a cavity as possible, i.e., when E is a relatively heavy element such as bismuth or lead. However, if the E_{10} cavity is too small, particularly when E is arsenic or germanium, then maximizing the internal volume of the polyhedral cavity can override any electronic considerations from the Wade–Mingos rules.

This consideration also applies to 8-vertex clusters with interstitial atoms. The most spherical 8-vertex deltahedron, namely the bisdisphenoid (Fig. 1), appears to have too small a cavity for an interstitial transition metal. However, the square antiprism has two fewer edges and can be partially flattened to make a puckered eight-membered ring, which can accommodate a transition metal in the center (Fig. 8). Known clusters of this type include $M@E_8{}^{n-}$ (M = Cr [98], Mo [98], Nb [99]; E = As, Sb; $n = 2, 3$ for Cr and Mo; $n = 3$ for Nb). The transition metal in such structures can be considered to be eight-coordinate with flattened square antiprismatic coordination. The $E_8{}^{8-}$ ring (E = As, Sb) can be considered formally to be an octaanion, isoelectronic with the common form of elemental sulfur, S_8. Thus in $M@E_8{}^{2-}$ (M = Cr, Mo; E = As, Sb), the central transition metal has the d^0 formal oxidation state of +6. Similarly in $Nb@E_8{}^{3-}$, the central niobium atom has its d^0 formal oxidation state of +5.

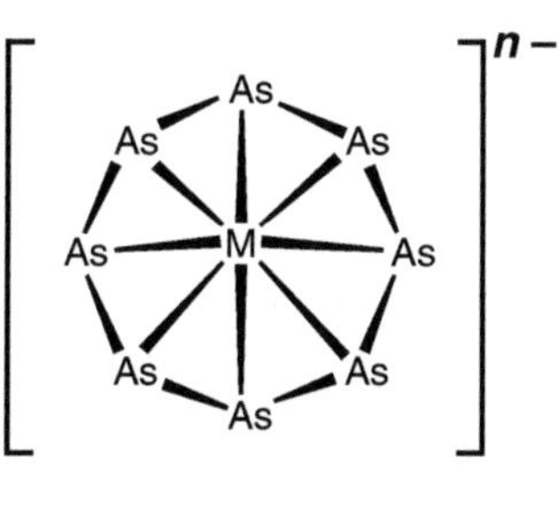

Fig. 8 The puckered eight-membered As_8 ring in the $MAs_8{}^{n-}$ derivatives. This puckered ring arises from severe flattening of a square antiprism

6 A Cluster Within a Cluster: The Icosahedral Matryoshka Nesting Doll Cluster

In addition to the possibility of a cluster containing a single interstitial atom as discussed above, there is the possibility of having a smaller cluster inside a larger cluster. Such a cluster has a structure resembling a Russian matryoshka nesting doll.

The unique example of a stable well-characterized post-transition element cluster of this type is the nickel–arsenic cluster $[As@Ni_{12}@As_{20}]^{3-}$, synthesized as a stable compound and characterized structurally by X-ray diffraction [100]. This cluster is also remarkable in exhibiting icosahedral symmetry throughout the entire nested structure. The outer layer of this cluster (Fig. 9) consists of a regular As_{20} dodecahedron. This polyhedron has a relatively large volume for an As_{20} cluster since it is highly symmetrical and has a minimum number of edges for a closed polyhedron. Removing this outer As_{20} dodecahedron next gives a Ni_{12} icosahedron. A single arsenic atom is located in the center of this Ni_{12} icosahedron.

The available electrons in this cluster can be allocated among the layers to give each layer the closed shell configuration [101]. Thus, allocating the overall –3 charge on the $[As@Ni_{12}@As_{20}]^{3-}$cluster to the central arsenic atom gives the As^{3-} anion, which has the stable configuration of the next noble gas, krypton. Next, consider the outer As_{20} dodecahedral shell, which has a total of $20 \times 5 = 100$ total valence electrons. These 100 electrons are partitioned into the 60 electrons required for two-center two-electron bonds along each of the 30 edges of the regular dodecahedron and the 40 electrons required for a lone pair on each of the 20 arsenic atoms. Extended Hückel calculations [100] indicate negligible interactions between the molecular orbitals defining the As–As bonds in the outer As_{20} dodecahedron and the molecular orbitals in the inner $As@Ni_{12}$ fragment.

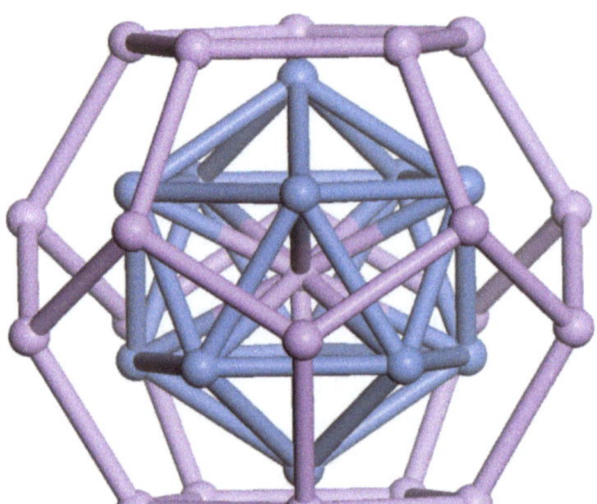

Fig. 9 The Ni_{12} icosahedron inside the As_{20} dodecahedron in the icosahedrally symmetric $[As@Ni_{12}@As_{20}]^{3-}$ cluster

Finally consider the intermediate Ni_{12} icosahedron layer. Bare nickel atoms in clusters are zero electron donors since they have filled d^{10} shells and in that sense function as pseudonoble gases [97]. However, the Ni_{12} icosahedron is well positioned to receive the 40 electrons from the 20 lone pairs of the arsenic atoms in the outer As_{20} dodecahedron. This gives the Ni_{12} cluster the favorable 40 total valence electron configuration suggested by the jellium model discussed above [21]. Note also that the lone pairs on each arsenic vertex of the As_{20} dodecahedron are not external lone pairs as might have been initially assumed. Instead they are directed toward the midpoints of the 20 triangular faces of the Ni_{20} icosahedron.

7 Summary

The Zintl ions and related bare post-transition metal clusters, first observed more than a century ago [1, 3–7], exhibit many interesting structural features that have provided challenges to theoretical chemists since the publication of the initial Zintl–Klemm concept [10, 11] and Mooser–Pearson rules [12] more than a half century ago. After definitive structural information first became available from Corbett's research group [8, 9], these initially used theoretical models soon proved to be inadequate. The resemblance of many of the initially determined post-transition metal clusters to those of the polyhedral boranes suggested extension of the Wade–Mingos rules [13–16] developed for borane chemistry, to such clusters. However, limitations of the Wade–Mingos rules were suggested by the unusual hypoelectronic cluster polyhedra of the group 13 metals found in intermetallics.

In general, the methods used to synthesize Zintl ions and related clusters are not very selective in most cases, since they cannot yet be designed in advance to give a cluster of a specific size or skeletal electron count. Thus, the synthesis of the Zintl cluster anions involves simply reducing the post-transition element with an alkali metal under suitably chosen conditions, normally involving reactions at high temperatures. The related cationic clusters, such as the $Pd@Bi_{10}{}^{4+}$ cation in $Bi_{14}PdBr_{16}$, are typically obtained by reduction of a halide of the post-transition metal in a strongly Lewis acidic medium. For this reason, the availability of reliable theoretical methods in recent years, particular those based on DFT, has allowed exploration of a variety of new cluster types in advance of the discovery of a method for their actual synthesis in the laboratory. The results from studies using such modern theoretical methods have led to significant advances in understanding the structure and bonding in bare post-transition metal clusters. Thus, theoretical prediction of different polyhedra for the isovalent $Ge_n{}^{2-}$ and $B_nH_n{}^{2-}$ derivatives for $n = 8$, 11, and 14 was critical in suggesting the need to develop theoretical models beyond the Wade–Mingos rules [21]. This paved the way for the observation that the jellium model, originally developed by physicists [22, 23], provides a useful rationalization of the special abundance of clusters with 20 total electrons, such as the P_4 known since chemical antiquity, as well as those with 40 total electrons. The latter include structures found in particularly diverse physical regimes such as the Zintl ion [8] Ge_9^{4-} found in liquid ammonia or in cryptate-complexed alkali metal "salts," the ion In_{11}^{7-} found in the K_8In_{11} intermetallic [66], and the ion Al_{13}^- found in gas-phase studies [74].

The chemistry of bare metal clusters with interstitial atoms is clearly the new frontier as indicated by two recent serendipitous discoveries:

1. The synthesis and structural characterization as a stable compound of the "cluster within a cluster" matryoshka doll anion [100] $[As@Ni_{12}@As_{20}]^{3-}$ with perfect icosahedral symmetry.
2. The even more recent synthesis and structural characterization of the metal-centered pentagonal prismatic clusters $M@Ge_{10}^{3-}$ (M = Fe [24], Co [25]) not predicted by any of the bonding theories.

These exciting recent developments are indications that much more new and exciting bare metal cluster chemistry remains to be discovered.

Acknowledgment The author is indebted to the US National Science Foundation for support of this work under Grant CHE-0716718. This article is dedicated to the memory of Prof. Ioan Silaghi-Dumitrescu (1950–2009) in recognition of the pivotal role that he played in density functional theory studies on bare post-transition element clusters.

References

1. Johannis A (1891) Compt Rendu Hebd Séances Acad Sci 13:795
2. Johannis A (1892) Compt Rendu Hebd Séances Acad Sci 114:587
3. Smyth FHJ (1917) Am Chem Soc 39:1299
4. Zintl E, Goubeau J, Dullenkopf W (1931) Z Phys Chem Abst A 154:1
5. Zintl E, Harder A (1931) Z Phys Chem Abst A 154:47
6. Zintl E, Dullenkopf W (1932) Z Phys Chem Abst B 16:183
7. Zintl E, Kaiser H (1933) Z Anorg Allgem Chem 211:113
8. Corbett JD (1985) Chem Rev 85:383
9. Corbett JD (1976) Prog Inorg Chem 21:129
10. Zintl E (1939) Angew Chem 52:1
11. Klemm W (1958) Proc Chem Soc (Lond) 329
12. Mooser E, Pearson WB (1956) Phys Rev 101:1608
13. Wade K (1971) Chem Commun 792
14. Wade K (1976) Adv Inorg Chem Radiochem 18:1
15. Mingos DMP (1984) Acc Chem Res 17:311
16. Mingos DMP, Johnston RL (1987) Struct Bond 68:29
17. King RB, Rouvray DH (1977) J Am Chem Soc 99:7834
18. King RB, Silaghi-Dumitrescu I, Lupan A (2005) Dalton Trans 1858
19. King RB, Silaghi-Dumitrescu I, Lupan A (2005) Inorg Chem 44:3579
20. King RB, Silaghi-Dumitrescu I, Uţă M (2008) Eur J Inorg Chem 3996
21. King RB, Silaghi-Dumitrescu I (2008) Dalton Trans 6083
22. De Heer WA (1993) Rev Mod Phys 65:611
23. Janssens E, Neukermans S, Lievens P (2004) Curr Opin Solid State Mater Sci 8:185
24. Zhou B, Denning MS, Kays DL, Goicoechea JM (2009) J Am Chem Soc 132:2802

25. Wang J-Q, Stegmaier S, Fässler TF (2009) Angew Chem Int Ed 48:1998
26. Hansen DA, Smith JF (1967) Acta Crystallogr 22:836
27. Zintl E, Dullenkopf W (1932) Z Phys Chem B16:195
28. Dong Z-C, Corbett JD (1996) Inorg Chem 35:2301
29. Williams RE (1971) Inorg Chem 10:210
30. Williams RE (1992) Chem Rev 92:177
31. King RB, Duijvestijn AJW (1990) Inorg Chim Acta 178:55
32. Aihara J (1978) J Am Chem Soc 100:3339
33. King RB (ed) (1983) Chemical applications of topology and graph theory. Elsevier, Amsterdam, pp 99–123
34. King RB (1976) In Liebman JF, Greenberg A (eds) Molecular structure and energetics. VCH, Deerfield Beach, Florida, pp 123–148
35. Stone AJ (1981) Inorg Chem 20:563
36. Stone AJ (1980) Mol Phys 41:1339
37. Stone AJ, Alderton MJ (1982) Inorg Chem 21:2297
38. Stone AJ (1984) Polyhedron 3:1299
39. Hirsch A, Chen Z, Jiao H (2001) Angew Chem Int Ed 40:2834
40. King RB (2001) Chem Phys Lett 338:237
41. Jemmis ED, Schleyer PvR (1982) J Am Chem Soc 104:4781
42. Schleyer PvR, Maerker C, Dransfeld A, Jiao H, van Eikema Hommes NJR (1996) J Am Chem Soc 118:6317
43. Kutzelnigg W, Fleischer U, Schindler M (1991) In: Diehl P, Fluck E, Gunter H, Kosfeld R, Seelig J (eds) NMR basic principles and progress, vol. 23. Springer, New York
44. Schleyer PvR, Jiao H, van Eikema Hommes NJR, Malkin VG, Malkina O (1997) J Am Chem Soc 119:12669
45. Corminboeuf C, Heine T, Weber J (2003) Phys Chem Chem Phys 5:246
46. Bohmann J, Weinhold FA, Farrar TC (1997) J Chem Phys 107:1173
47. Wannere CS, Schleyer PvR (2003) Org Lett 5:605
48. Schleyer PvR, Subramanian G, Jiao J, Najafian K, Hofmann M (1997) In: Siebert W (ed) Advances in boron chemistry. The Royal Society of Chemistry, Cambridge, England, pp 3–14
49. Schleyer PvR, Najafian K (1998) In: Casanova J (ed) The borane, carborane and carbocation continuum. Wiley, New York, p 169
50. Schleyer PvR, Najafian K (1998) Inorg Chem 37:3455
51. Najafian K, Tidwell TT, Schleyer PvR (2003) Inorg Chem 42:4190
52. Burns RC, Corbett JD (1981) J Am Chem Soc 103:2627
53. Guggenberger LJ (1968) Inorg Chem 7:2260
54. King RB (1982) Inorg Chim Acta 57:79
55. Wahl B, Ruck M (2010) Z Anorg Allgem Chem 636:337
56. Spiekermann A, Hoffmann SD, Fässler TF (2008) Angew Chem Int Ed 45:345
57. Dobrott RD, Lipscomb WN (1962) J Chem Phys 37:1779
58. King RB, Heine T, Corminboeuf C, Schleyer PvR (2004) J Am Chem Soc 126:430
59. Corminboeuf C, King RB, Schleyer PvR (2007) ChemPhysChem 8:391
60. Chen Z, Neukemans S, Wang X, Janssens E, Zhou Z, Silverans RE, King RB, Schleyer PvR, Lievens P (2006) J Am Chem Soc 126:12829
61. Cui L-F, Huang X, Wang LM, Zubarev DY, Boldyrev AI, Li J, Wang LS (2006) J Am Chem Soc 126:8390
62. Cui L-F, Huang X, Wang LM, Li J, Wang LS (2006) J Phys Chem A 110:10169
63. Esenturk EN, Fettinger J, Eichhorn B (2008) J Am Chem Soc 128:9175
64. Corbett JD (1987) Struct Bond 87:157
65. King RB (1996) Inorg Chim Acta 252:115
66. Sevov SC, Corbett JD (1991) Inorg Chem 30:4877
67. Henning RW, Corbett JD (1997) Inorg Chem 36:6045

68. Dong Z-C, Corbett JD (1995) J Cluster Sci 6:871
69. King RB, Silaghi-Dumitrescu I, Kun A (2002) Dalton Trans 3999
70. King RB, Silaghi-Dumitrescu I (2003) Inorg Chem 42:6701
71. King RB, Silaghi-Dumitrescu I, Uţă MM (2006) Inorg Chem 45:4974
72. King RB, Silaghi-Dumitrescu I, Uţă MM (2007) Dalton Trans 364
73. Dong L, Chen HS, Xie Z (2005) Angew Chem Int Ed Engl 44:2128
74. Bergeron DE, Castleman AW Jr, Morisato T, Khanna SN (2004) Science 304:84
75. Madelung E (1936) Die mathematischen Hilfsmittel der Physikers. Springer, Berlin
76. Demkov YuN, Ostrovskii VN (1972) Sov Phys JETP 35:66
77. Purdela D (1988) Int J Quant Chem 34:107
78. Klechkovskii VM (1951) Dokl Akad Nauk 86:691
79. Chen Z, King RB (2005) Chem Rev 105:3613
80. Queneau V, Sevov SC (1997) Angew Chem Int Ed 36:1754
81. Corbett JD (1997) Struct Bond 87:157
82. Corbett JD (2000) Angew Chem Int Ed 39:670
83. Fässler TF, Hoffmann SD (2004) Angew Chem Int Ed 43:6242
84. Scharfe S, Fässler TF, Stegmaier S, Hoffmann SD, Ruhland K (2008) Chem Eur J 14:4479
85. Esenturk EN, Fettinger J, Eichhorn B (2005) Chem Commun 247
86. Esenturk EN, Fettinger J, Lam Y-F, Eichhorn B (2004) Angew Chem Int Ed 43:2132
87. Wang J-Q, Stegmaier S, Wahl B, Fässler TF (2010) Chem Eur J 16:1793
88. Goicoechea JM, Sevov SC (2005) J Am Chem Soc 127:7676
89. Kocak FS, Zavalij P, Lam Y-F, Eichhorn BW (2008) Inorg Chem 47:3515
90. Esenturk EN, Fettinger JC, Eichhorn BW (2006) J Am Chem Soc 126:12
91. Henning RW, Corbett JD (1999) Inorg Chem 38:3883
92. Sevov SC, Corbett JC (1993) Inorg Chem 32:1059
93. Ruck M, Dubenskyy V, Söhnel T (2003) Angew Chem Int Ed 45:2978
94. King RB, Silaghi-Dumitrescu I, Uţă MM (2008) Chem Eur J 14:4542
95. King RB, Silaghi-Dumitrescu I, Uţă MM (2009) J Phys Chem A 113:527
96. King RB, Silaghi-Dumitrescu I, Uţă M (2009) Inorg Chem 48:8508
97. King RB (2004) Dalton Trans 3420
98. Kesanli B, Fettinger J, Eichhorn B (2003) J Am Chem Soc 125:7367
99. Kesanli B, Fettinger J, Eichhorn B (2004) Inorg Chem 43:3840
100. Moses MJ, Fettinger JC, Eichhorn BW (2003) Science 300:778
101. King RB, Zhao J (2006) Chem Commun 4205

Struct Bond (2011) 140: 25–57
DOI: 10.1007/430_2011_43

Published online: 12 May 2011

Polyanions of Group 14 and Group 15 Elements in Alkali and Alkaline Earth Metal Solid State Compounds and Solvate Structures

Stefanie Gärtner and Nikolaus Korber

Abstract The chapter discusses the formation of discrete polyanions of group 14 and group 15 elements in solid state structures of alkali and alkaline earth metal compounds depending on the degree of reduction and the cation radii. Applying this concept reveals areas of stability of polyanions of different nature. The mapping of the compounds in one diagram presents a comprehensive overview for the regions of existence of the polyanions in the discussed materials. Additionally, solvate structures are summarized, which provide a promising possibility to adjoin new polyanionic arrangements.

Keywords Cation radii · Group 14 · Group 15 · Homoatomic polyanions · Solid state · Solution · Zintl

Contents

S. Gärtner and N. Korber (✉)
Institute for Inorganic Chemistry, University of Regensburg, D-93040 Regensburg, Germany
e-mail: nikolaus.korber@chemie.uni-regensburg.de

1 Introduction

The existence of polyanions is well known since Joannis first observed the change of color of the characteristic blue sodium ammonia solutions, which turned green in the presence of elemental lead [1, 2]. This was due to the formation of enneaplumbide clusters in solution, and one of the first experimental hints for the availability of metals in negative oxidation states. In the 1930s, Zintl and coworkers were able to identify the compositions of a large number of compounds with negatively charged main group metals and semi-metals using potentiometric titration methods in alkali metal liquid ammonia solutions [3, 4]. This low temperature route of reducing elements of group 14 and group 15 in solution is limited to the temperature range of liquid ammonia between −33 and −77°C. Higher reduced and thus more electron-rich solid state compounds can be obtained by reacting electropositive elements with elements beyond the *Zintl* border in solid state reactions at high temperatures. This route is especially important for elements that cannot be directly reduced at low temperatures, as is the case for silicon or germanium. From the last century until now, a large and still growing number of so-called *Zintl* phases have been reported for alkali or alkaline earth metals in combination with heavier homologues of group 14 and group 15. They all contain polyanionic partial structures of (semi)metal, which range from three-dimensionally extended networks and layers in two dimensions to infinite chains and molecular, discrete polyanions, depending on the degree of reduction. These polyanions feature homopolar element–element bonds exclusively and are connected by electrostatic interactions with the ions of the electropositive metal; the compounds can therefore be understood as polyanionic salts. Using polar solvents which are resilient against reduction and which are able to stabilize high charges, some of these polyanions can be extracted into solution. Liquid ammonia or ethylenediamine can be used as solvents. In solution, polyanions become more and more popular as molecular building blocks for further chemical transformations, e.g., reaction with transition metal complexes [5] or oxidation to bigger aggregates [6–9] and even new elemental modifications [10]. Besides their growing role in solution chemistry, polyanions continue to be the focus of interest in solid state sciences due to their interesting electronic properties and the possibility of studying the transition from (semi)metallic to ionic structures. It became evident that the use of different or different combinations of electropositive metals led to variations in size and shape of the polyanionic moiety. For some elements this was extensively studied, and a direct influence of cation-element interaction on size, structure, and functionality of the polyanions was emphasized. These interactions are naturally strong for small cations and become less strong the higher the period, in accord with the corresponding radii.

The second important variable for size, structure, and functionality is the degree of reduction of the compound. In *Zintl* phases, there exists a well-defined relationship between chemical and electronic structures [11]. Assuming complete electron transfer from the electropositive to the electronegative metal restates the classical valence rules by giving an average number of valence electrons per anion atom (1). Combined with the 8-N rule, this gives the average number of E–E contacts b_e that occur in the structure (2). Zintl's proposal led to the pseudoatom concept for the charged unit, introduced by Klemm, which relates the structural characteristics of the polyanion to the isoelectronic element structure. The combination of these approaches is well established as the *Zintl–Klemm(–Busmann)* concept [11, 12].

$$N_E = (m \cdot n_M + e \cdot n_E)/e = m/e \cdot n_M + n_E, \quad (1)$$

$$b_e = 8 - N_E, \quad (2)$$

where N_E is the valence electron number per anion atom; m the number of valence electrons of the electropositive element, $m = 1$ for alkali metal, $m = 2$ for alkaline earth metal; n_M the number of electropositive atoms in sum formula; e the number of valence electrons of the electronegative element, $e = 4$ for group 14 element, $e = 5$ for group 15 element; n_E the number of electronegative atoms in sum formula; b_e the average number of E–E contacts.

In the early 1970s, Parthé introduced the valence electron concentration as a powerful systematic tool to explain and predict tetrahedral structures [13]. This concept made it possible to understand structural aspects by only knowing the chemical formula. It has to be noted that the complete electron transfer has to be handled with care, especially for small cations and lighter main group elements, due to not exclusively ionic interactions in this case. The bonding situation of various compounds has been discussed elsewhere [11, 14, 15] and will not be matter of detailed discussion in the overview presented here. We rather concentrate on a systematic consideration of discrete polyanions of group 14 and group 15 elements in solid state compounds. Therefore, the different homoatomic group 14 and group 15 polyanions are summarized and their structural properties will be briefly presented. Because it is of interest to have formalisms by which structural features can be expressed in terms of the chemical formula of the compound, we use the valence electron number per anion atom N_E as an average degree of reduction of the compound. Additionally, the (average) ionic radii of the involved electropositive element cations are taken into account. The scope of this chapter is to give an overview of the known homoatomic molecular polyanions in alkali and alkaline earth metal solid state compounds of group 14 and group 15 elements to find regions of stability for certain polyanions with respect to the average degree of reduction and the counter ion size. In the end of the chapters of solid state compounds, a map of the differently natured polyanions in certain (N_E; $r(M^{I/II})_{av}$) areas is provided.

Some of the homoatomic polyanions formulated for solid state compounds are also stable in solution, and even new polyanions are accessible by crystallization from these approaches. The final chapter summarizes homoatomic polyanions of group 14 and group 15, which have been reported in solvate structures.

2 Homoatomic Polyanions of Group 14 Elements in Solid State Compounds

2.1 Clusters of Group 14 Elements in Compounds with 4.4 ≤ N_E ≤ 5

The rather low degree of reduction between 4.4 and 5.5 e/atom implicates that atoms of group 14 are bound to three or more bonding partners in average. Coming from the fourfold-bound atoms in the elemental structures ($N_E = 4$), isolated polyanions compete with extended anionic partial structures in compounds, which of course also exhibit these low N_E values close to $N_E = 4$. The most common polyanions of group 14 elements are homoatomic clusters, in particular the tetrahedrally shaped E_4^{4-} and the distorted monocapped square antiprismatically shaped E_9^{4-} clusters [5]. The nine-atom clusters are found in binary alkali metal compounds with the composition $M_4^IE_9$ and $M_{12}^IE_{17}$. Table 1 gives an overview of the hitherto known E_9^{4-}-containing compounds. Calculating the valence electron per anion atom, N_E shows the nonatetrelide clusters being part of crystal structures with $N_E = 4.4$–4.7 e/atom. With an average cation radius between 0.133 and 0.167 nm [27], compounds containing these nine-atom clusters tend to be stabilized in crystal structures by rather large and therefore polarizable counter ions. This is in accordance with the HSAB principle (for a recent review on HSAB principle, see [28]). $Na_{12}Ge_{17}$ represents an exception; the crystallization of nonagermanide cages together with small sodium cations ($r_{Na+} = 0.098$ nm) results in huge unit cell dimensions and very poor intensity data [25]. This gives an idea about the still missing crystal structures of small cations and nonatetrelide cage anions. Presumably, these phases do not form at all or do not form suitable crystalline products for X-ray structure analysis.

The second most common cluster species are represented by the tetrahedrally shaped fourfold negatively charged E_4^{4-} cages, which can be found together with E_9^{4-} anions in $M_{12}^IE_{17}$ compounds. Additionally, they represent the single anionic moiety in $M_4^IE_4$ and $M^{II}E_2$ materials with a N_E value of 5, corresponding to threefold bound group 14 atoms. The average radius of the cations in this sort of compounds ranges from 0.098 to 0.167 nm and covers a wider area compared to the

Table 1 Solid state compounds that contain nonatetrelide clusters

		Compound	N_E	$r(M^I)_{av}$	References
E_9^{4-}	Si	–			
	Ge	K_4Ge_9	4.4	0.133	[16]
		Cs_4Ge_9	4.4	0.167	[17]
	Sn	K_4Sn_9	4.4	0.133	[18]
	Pb	K_4Pb_9	4.4	0.133	[19]
		Rb_4Pb_9	4.4	0.148	[20]
		Cs_4Pb_9	4.4	0.167	[21]
		$Cs_{10}K_6Pb_{36}$	4.4	0.154	[20]
E_9^{4-}/E_4^{4-} (6:7)	Sn	$K_{52}Sn_{82}$	4.63	0.133	[22]
		$Cs_{52}Sn_{82}$	4.63	0.167	[22]
E_9^{4-}/E_4^{4-} (1:2)	Si	$K_{12}Si_{17}$	4.7	0.133	[23]
		$Rb_{12}Si_{17}$	4.7	0.148	[24]
	Ge	$Na_{12}Ge_{17}$	4.7	0.098	[25]
		$K_{12}Ge_{17}$	4.7	0.133	[26]
	Sn	$K_{12}Sn_{17}$	4.7	0.133	[23]
		$Rb_{12}Sn_{17}$	4.7	0.148	[23]

nine-atom species discussed above. Below $N_E = 0.098$, only three-dimensionally extended network structures are observed. But as mentioned before, the compounds with molecular anions compete with those with extended networks, and the kind of observed anionic moiety is sensitive toward the pressure applied during synthesis [29–31] (Table 2).

In addition to these well-known clusters, a third interesting cluster family has been described by Sevov et al. [50]. In ternary compounds with the composition $M_4^ILi_2Sn_8$ (A = K or Rb), a square antiprismatically shaped Sn_8 cluster is present. The valence electron number per tin atom equals 4.75 and lies between the values for nine- and four-atom clusters. The use of small lithium combined with the larger potassium or rubidium ions leads to the formation of square antiprisms Sn_8^{6-}, in which the two square faces are capped by lithium cations, and two edges of the cluster are additionally spanned by lithium cations. The average radius adds up to 0.111 (M^I = K) and 0.121 (M^I = Rb). The authors give two approaches to the interpretation of the shape of the cluster. Assuming complete electron transfer, the Sn_8 cluster is sixfold negatively charged and therefore an *arachno* species according to Wade's rules [51–53]. The second approach sees lithium interacting partial covalently, which results in a fourfold negatively charged ten-atom *closo*-cluster (Fig. 1).

This Sn_8 cluster is also found in the compound $Ba_{16}Na_{204}Sn_{310}$ [54], where it is only coordinated by sodium cations. The number of valence electrons per atom of 4.76 in this compound is only slightly different from the one in the $Li_2M_4^ISn_8$ phases, and the average radius of the cations lies with 0.101 nm in the same range as well. This supports the introduced formalism of anion characteristic [N_E; $r(M)_{av}$] tuples. Beside the Sn_8 cages, $Ba_{16}Na_{204}Sn_{310}$ also contains gigantic clusters of 56 tin atoms with a charge of −44. They are endohedrally stabilized by four Ba^{2+} cations. The third cluster in this complex compound is represented by interstitially sodium-stabilized Sn_{16-n} clusters with the shape of a capped truncated tetrahedron (*Frank–Kasper* polyhedron) (Fig. 2).

Table 2 Solid state compounds that contain E_4^{4-} clusters, in addition to those mentioned in Table 1

		Compound	N_E	$r(M^{I/II})_{av}$	References
E_4^{4-}	Si	Na_4Si_4	5	0.098	[32, 33]
		K_4Si_4	5	0.133	[34, 35]
		Rb_4Si_4	5	0.148	[34, 35]
		Cs_4Si_4	5	0.167	[34, 35]
		K_7LiSi_8	5	0.125	[36]
		K_3LiSi_4	5	0.117	[36]
		$BaSi_2$	5	0.143	[37]
		$Sr_{1-x}Ba_xSi_2$ $x = 0.2–1.0$	5	0.130	[38]
	Ge	Na_4Ge_4	5	0.098	[32]
		K_4Ge_4	5	0.133	[34, 39]
		Rb_4Ge_4	5	0.148	[34, 39]
		Cs_4Ge_4	5	0.167	[34, 39]
		Rb_7NaGe_8	5	0.142	[40]
		$Na_2Cs_2Ge_4$	5	0.133	[41]
		K_7NaGe_8	5	0.129	[40]
		$SrGe_2$	5	0.127	[42]
		$BaGe_2$	5	0.143	[42]
	Sn	Na_4Sn_4	5	0.098	[43, 44]
		K_4Sn_4	5	0.133	[44, 45]
		Rb_4Sn_4	5	0.148	[46]
		Cs_4Sn_4	5	0.167	[46]
	Pb	Na_4Pb_4	5	0.098	[47]
		K_4Pb_4	5	0.133	[48]
		Rb_4Pb_4	5	0.148	[49]
		K_3LiPb_4	5	0.117	[20]
		Rb_3LiPb_4	5	0.128	[20]
		Cs_4Pb_4	5	0.167	[49]
		Cs_3LiPb_4	5	0.142	[20]
		Cs_3NaPb_4	5	0.149	[20]

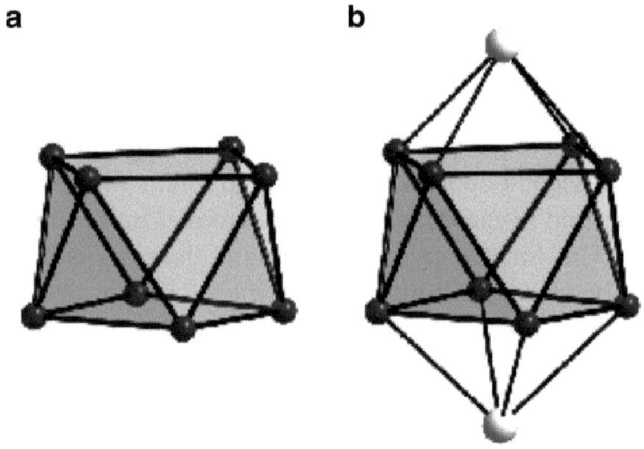

Fig. 1 (**a**) *Arachno*-$[Sn_8^{6-}]$; (**b**) *closo*-$[Li_2Sn_8]^{4-}$ cluster in $M_4^ILi_2Sn_8$

Obviously, it is possible to produce bigger clusters at $N_E = 4.76$ by stabilizing them with suitably sized cations, available by electroequivalent substitution of alkali metal by alkaline earth metal cations.

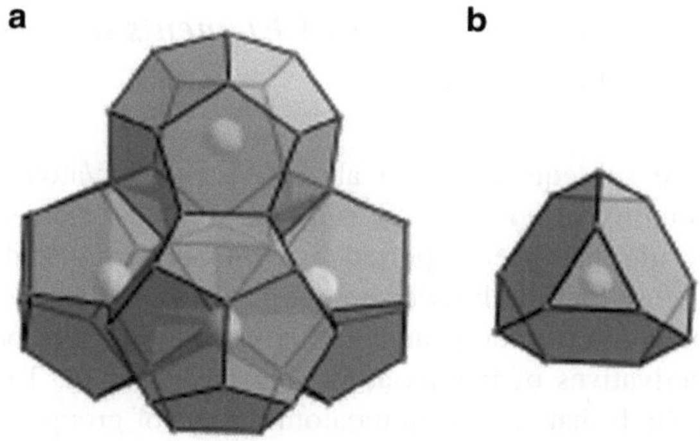

Fig. 2 (**a**) Sn_{56}^{44-} cluster filled with four Ba^{2+} cations in $Ba_{16}Na_{204}Sn_{310}$; (**b**) interstitially metal stabilized $M@E_{12}^{12-}$ capped truncated tetrahedra in $Ba_{16}Na_{204}Sn_{310}$ (E = Sn, M = Na), $M^{II}Na_{10}Sn_{12}$ (E = Sn, M = Ca, Sr), and Li_7RbGe_8 (E = Ge, M = Li)

The concept of mixed alkali/alkaline earth metal cations made it also possible to generate compounds of tin with $N_E = 5$ and $r(M)_{av} < 0.098$ nm in $CaNa_{10}Sn_{12}$ and $SrNa_{10}Sn_{12}$ [55]. The cluster species one finds in these compounds are again capped truncated tetrahedra, which are stabilized by the endohedral M^{II} cations calcium and strontium. Analogous germanium cages are observed in Li_7RbGe_8 ($N_E = 5$, $r(M)_{av} = 0.078$ nm), where lithium is interstitially bound. Additional Ge_4^{4-} tetrahedra, which possess the same N_E value, can be found in the crystal structure [56].

2.2 *In Between Group 14 Clusters and Rings:* $N_E = 5.5$

According to the *Zintl–Klemm* concept, an increased degree of reduction of $N_E = 5.5$ means that one atom is in average bound to 2.5 neighbors. This N_E value represents the border between clusters ($N_E < 5$) and rings ($N_E > 5.6$), of course competing with two-dimensionally extended layers and one-dimensionally extended strands. In the binary system Ba–Si, this can be well demonstrated with the help of Ba_3Si_4 [57, 58]. This compound contains anions shaped like a butterfly with the formal charge distribution $[Ba^{2+}]_3[(3b)Si^{1-}]_2[(2b)Si^{2-}]_2$. Two silicon atoms are threefold bound like in the previously introduced E_4^{4-} tetrahedra. The remaining two silicon atoms only possess two bonding partners, as one would expect in rings, and they are assigned a charge of −2 in accordance with the 8-N rule. In other words, due to the two-electron reduction of the Si_4^{4-} tetrahedra, one Si–Si bond is broken yielding a butterfly shaped Si_4^{6-} anion. The peritectic temperature of Ba_3Si_4 was determined to be 1,307(5) K, and the material is stable at room temperature [58]. The case is different in the Ba-Ge system, where two modifications of Ba_3Ge_4 are found [59]. The room temperature phase α-Ba_3Ge_4 crystallizes in a new structure type which is not isotypic but related to Ba_3Si_4 with one-dimensional chains built of distorted Ge_4 rings. At temperatures above 630 K, a phase is detected which is isotypic to Ba_3Si_4 with quasi-isolated Ge_4^{6-} anions.

2.3 Rings of Group 14 Elements in Compounds with $5.66 \leq N_E \leq 6$

The subsequent region above $N_E = 5.5$ e/atom comprises the narrow area of stability of homoatomic rings of group 14 elements. For silicon, the majority of results has been reported by von Schnering and Nesper, whereas Sevov et al. focused on the heavier homologues tin and lead. The number of ring constitutive atoms in all compounds either sums up to five or six, and for silicon additional derivatives of the six-atom ring are reported. Table 3 contains solid state compounds that contain homoatomic rings of group 14 elements.

The lithium and five-atom ring containing compounds ($Li_{12}Si_7$, Li_8MgSi_6, Ge_6Li_{11}) all show ecliptically stacked quasi-aromatic E_5 rings, where lithium is inserted between adjacent rings. This was called an *infinite sandwich complex* [62]. The same structural motif is found in Na_8BaSn_6 and Na_8BaPb_6, where contrary to the silicon or germanium compounds the infinite sandwich is built by coordination of the ring plane by the alkaline earth metal. The interchanged positioning concerning alkali/alkaline earth metal was attributed to the different size of the cations, and the ring plane seems to be preferably coordinated by the larger provided cation [65]. This holds true as well for the structures of Li_5CaSn_{11} and $Li_{8.84}CaSn_{6.16}$, where calcium acts as bridging atom. These observations support the major influence of size with respect to the position of cations toward the anion.

A second remarkable observation is that all crystal structures which contain five-atom rings additionally include higher reduced anionic species. These auxiliary anions seem to be obligatory for the stabilization of the ring structures and become intelligible by taking the N_E value into account. The N_E value of pure quasi-aromatic E_5^{6-} rings sums up to 5.2, which is quite close to the value for homoatomic clusters. At this N_E value, ring formation is not favored due to the remarkable stability of the clusters. Therefore, the ring-containing compounds rather fill the gap between rings and chains at N_E values between 5.6 and 6, but to achieve this high degree of reduction, further and higher reduced anionic species are compulsory in

Table 3 Reported homoatomic rings in alkali/alkaline earth metal containing solid state compounds of group 14 elements

		Compound	Add. anion	N_E	$r(M^{I/II})_{av}$	References
E_5 ring	Si	$Li_{12}Si_7$	Si_4^{12-}	5.71	0.068	[60, 61]
		Li_8MgSi_6	Si^{4-}	5.66	0.069	[62]
	Ge	Ge_6Li_{11}	Ge^{4-}	5.83	0.068	[63]
	Sn	Na_8BaSn_6	Sn^{4-}	5.66	0.103	[64]
		$Li_5Ca_7Sn_{11}$	Sn_6 chain	5.73	0.090	[65]
		$Li_{8.84}CaSn_{6.16}$	Sn^{4-}	5.76	0.072	[64]
	Pb	Na_8BaPb_6	Pb^{4-}	5.66	0.103	[65]
E_6^{10-} ring	Si	$Ba_4Li_2Si_6$	–	5.66	0.118	[66]
		$Sr_4Li_2Si_6$	–	5.66	0.107	[67]
	Ge	$Ba_4Li_2Ge_6$	–	5.66	0.118	[66]
Si_{12}^{21-}		$Ca_7Mg_{7.5+/-\delta}\,Si_{14}$	Si^{4-}	6.07	0.090	[68]
Si_{10}^{20-}		SiSr	–	6	0.127	[69]

the phases. For E_6^{10-} rings the case is different. The N_E values for these polyanions amounts to 5.66 without any additional anion in the crystal structure; therefore, these rings may be found as the exclusive anionic moiety in $M_4^{II}Li_2E_6$ compounds. The chemical bonding in these highly reduced polyanions is not trivial at all. In 1985, Corbett stated (...) *Such inexplicable bonding arrangements between the heavy elements boggle the mind!* (...) [70], and up to now this still is a fascinating area of research for experimental as well as theoretical chemists. The *Zintl–Klemm* and 8-N rule do not apply simply because of multiple bonding between the main group elements. The bonding in the E_6^{10-} 34-electron systems can be elucidated in terms of the Hückel formalism: 24 electrons form six σ bonds and six nonbonding electron pairs. Ten electrons are remaining and contribute to the π-system by filling three bonding and two antibonding MOs. Based on the total wave function, the electron localization function (ELF [71, 72]) was calculated, which indicated a formally aromatic character of the ring [66].

For silicon, two exotic ring structures have been reported. In $Ca_7Mg_{7.5+/-\delta}$ Si_{14}, six atoms form a ring, and each atom is substituted by an additional Si atom to form a planar Si_{12} group of D_{6h} symmetry, for which a charge of $-21 \pm 2\delta$ is reported due to not fully occupied magnesium positions. This is compatible neither with a benzene-like unit (Si_{12}^{18-}) nor with an exclusively single-bonded unit (Si_{12}^{24-}). The intraring bond distances indicate a partial double-bonding character because they are somewhat shorter than Si–Si single bonds, and according to the filled π-states for the planar system this means 1.75 double bonds are distributed over the whole six-membered ring [68].

For the binary alkaline earth silicide SiSr, two different structures have been reported. One contains one-dimensionally extended zigzag chains beside isolated Si^{4-} atoms [73]. Schäfer et al. prepared a modification with the same composition, which instead contained isolated silicide units of ten atoms. In these units, planar hexagons are substituted in the 1-, 2-, 4-, and 5-ring positions by four additional Si atoms. An isostructural compound was found for germanium as well, but showing defects in this unit in the positions 1, 2, 4, and 5. Both materials could not be obtained from stoichiometric approaches, and their formation obviously is coupled to strontium excess [69] (Fig. 3).

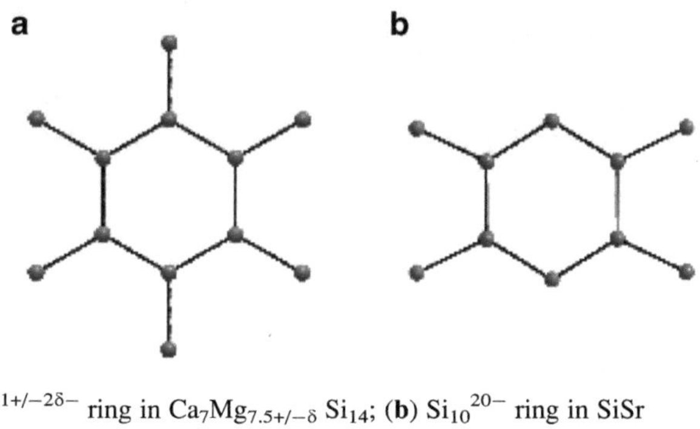

Fig. 3 (**a**) $Si_{12}^{(21+/-2\delta-}$ ring in $Ca_7Mg_{7.5+/-\delta}$ Si_{14}; (**b**) Si_{10}^{20-} ring in SiSr

2.4 E_n *Chains* ($n > 2$) *of Group 14 Elements in Compounds with* $6 < N_E \leq 6.6$

A further enhanced degree of reduction leads to chains of group 14 elements with N_E values between 6 and 6.6 e/atom. These N_E values cover the range between twofold bound silicon atoms in one-dimensionally extended zigzag chains and E_2^{6-} dumbbells. Isolated chains reside in between these border structures and therefore compete with them [14]. This leads to the expectation of an extreme cation size dependence of the anion formation. Table 4 contains the reported isolated homoatomic chains in solid state compounds.

By having a closer look at the N_E values and the average radii of the cations, it becomes evident that in contrast to the rings discussed above, the degree of reduction plays a subordinate role, within the given borders, for the constitution of the anions. For example, a four-atom chain is observed in $Ca_8Li_{0.969}Mg_{2.031}Si_8$ with an N_E value of 6.63 due to additional monoatomic Si^{4-} anions in the crystal structure. The same length of the chain is also present in $Ba_6Mg_{5.3}Li_{2.7}Si_{12}$ as the exclusive anionic moiety with a N_E value of 6.11. At nearly identical (N_E; $r(M^I/^{II})_{av}$) pair $Sr_{12}Mg_{17.8}Li_{2.2}Si_{20}$ was reported, which includes a three-atom chain and monoatomic Si^{4-} ions. This demonstrates the sensitive relationship between cation properties and anion formation. The majority of chains have up to now been observed within a very narrow range of averaged cation radii between 0.097 and 0.106. Two compounds have been reported for which the cation radius lies quite far away from this region. In $Sr_{11}Mg_2Si_{10}$, eight-atom chains could be realized with a corresponding averaged cation radius of $r(M^{I/II})_{av} = 0.119$ [14]. In monoclinic Li_7Sn_3, three-atom chains of tin are observed, which are exclusively coordinated by lithium. The arrangement of all atoms together in the unit cell is reminiscent of the structure of metallic lithium. Therefore, the structure of Li_7Sn_3 can be understood as variant of the cubic body-centered metal lattice [80].

Homoatomic chains are a fascinating group of polyanions, due to the possibility to study multiple bonding of heavier main group elements in the solid state and the coordination behavior of differently charged atoms in the same molecular structure.

Table 4 Homoatomic chains of group 14 elements in solid state compounds

	Chain length	Compound	Add. anion	N_E	$r(M^{I/II})_{av}$	References
Si	$n = 3$	$Sr_{12}Mg_{17.8}Li_{2.2}Si_{20}$	Si^{4-}	6.11	0.104	[74]
	$n = 6$	$Ba_2Mg_3Si_4$	Si_2^{6-}	6.5	0.104	[75]
		$Ba_6Mg_{5.3}Li_{2.7}Si_{12}$	–	6.11	0.104	[76]
		$Ca_8Li_{0.969}Mg_{2.031}Si_8$	Si^{4-}	6.63	0.097	[77]
	$n = 8$	$Sr_{11}Mg_2Si_{10}$	Si^{4-}	6.60	0.119	[14]
Ge	$n = 4$	Ca_7Ge_6	Ge_2^{6-}	6.33	0.106	[78]
	$n = 6$	$Ca_8Ge_8Li_{1.18}Mg_{1.82}$	Ge^{4-}	6.60	0.097	[77]
		$Ba_6Ge_{12}Li_{3.1}Mg_{4.9}$	–	6.07	0.103	[79]
Sn	$n = 3$	Li_7Sn_3	–	6.33	0.068	[80]
	$n = 4$	Ca_7Sn_6	Sn_2^{6-}	6.33	0.106	[81]
	$n = 6$	$Li_5Ca_7Sn_{11}$	Sn_5^{6-}	5.73	0.090	[65]

Table 5 Charge of group 14 chains in solid state compounds according to the chemical formula compared to the charge expected for *Zintl*-compliant σ-bonded chains

Chain	Charge according formula q	Charge according Zintl z	Compound
Si_3	−7.45	−8	$Sr_{12}Mg_{17.8}Li_{2.2}Si_{20}$
Si_6	−14	−14	$Ba_2Mg_3Si_4$
	−12.65	−14	$Ba_6Mg_{5.3}Li_{2.7}Si_{12}$
	−13.03	−14	$Ca_8Li_{0.969}Mg_{2.031}Si_8$
Si_8	−18	−18	$Sr_{11}Mg_2Si_{10}$
Ge_4	−8	−10	Ca_7Ge_6
Ge_6	−12.82	−14	$Ca_8Ge_8Li_{1.18}Mg_{1.82}$
	−12.45	−14	$Ba_6Ge_{12}Li_{3.1}Mg_{4.9}$
Sn_3	−7	−8	Li_7Sn_3
Sn_4	−8	−10	Ca_7Sn_6
Sn_6	−13	−14	$Li_5Ca_7Sn_{11}$

Nesper et al. intensively studied silicide chains in ternary and quaternary alkali and alkaline earth metal compounds [82]. It could be shown that the more highly charged terminating silicon atoms prefer coordination by small and hard Li and Mg atoms, whereas the less-charged Si atoms in between obviously tend to be coordinated by larger and soft cations [14, 15]. The charge distribution in the chains is not trivial, and multiple bonding between the heavy elements has to be taken into account. To obtain a first hint on the degree of reduction, one can count the formal charges of the chains from the chemical formula, adopting a charge of −4 for monoatomic anions, −6 for dumbbells, −8 for three-atom chains, and so on, and compare the result with the expected value according to the *Zintl–Klemm* concept which gives the value for single-bonded chains (Table 5).

The discrepancy between the charge according to the chemical formula q and the charge according to single-bonded atoms z gives the double bond contribution to the total bonding x as $x = z - q$, whereby x electrons are taken out of the anti-bonding π^* orbitals. Therefore, smaller bonding distances are expected which are observed more or less distinctively in the latter compounds. This also fits with the observed planarity and the near 120° bonding angles of the building blocks. For the case where $q = z$, one would expect only σ-bonded chains, but here also planar units and bonding angles near 120° are observed, which may mean that the electron pairs at the twofold bound atoms tend to feature sp^2 and p character due to orbital symmetry reasons [14].

Attention has to be drawn to the solid state compound $Li_5Ca_7Sn_{11}$, which includes Sn_5^{6-} rings (see Sect. 2.3) besides Sn_6 chains. All distances in the Sn_5 ring are shorter than single bonds. Calculations performed on one of the rings resulted in three filled π-bonding orbitals and two empty π-antibonding orbitals, which define a HOMO-LUMO gap of about 2.25 eV for Sn_5^{6-} [64]. The average distance within the five-membered ring is slightly elongated compared to an electronically balanced Na_8EuSn_6, but one distance is reported to be very short. Therefore, some degree of localization of the π-bonding is to be assumed. To the Sn_6 chain, a charge of −12 is assigned due to one significantly shorter Sn–Sn bond. The total amount of electrons from the electropositive metals in this material adds up to 19. Consequently, there is one electron left unassigned to the polyanions, which is in agreement with magnetic and conductivity measurements. The extra electron is reported to be delocalized over π-antibonding tin orbitals of the pentagon and the Sn_6^{12-} hexamers. Calcium *s* orbitals are also reported to be involved in the delocalization.

2.5 A Branched Four-Atom Tetrelide "Star" (E = Si, Ge)

Besides clusters, rings, and chains, a highly charged tetrelide anion that takes the shape of a three jagged star has been reported for silicon and germanium in different crystal structures. It was reported for the first time in 1980 by von Schnering et al. for silicon in the compound $Li_{12}Si_7$ [61], which additionally contains the before-mentioned Si_5^{6-} rings. In this compound, the four-atom anion is assigned a formal charge of −12, which is not understandable by the model of formal ions. A first interpretation was given by von Schnering et al., who suggested the formulation $(Li^+)_{24}(Si_4^{8-})(Si_5^{8-})_2$ with the Si_4^{8-} moiety isoelectronic to carbonate and the Si_5^{8-} to cyclopentene [61]. Leuken et al. carried out ab initio localized spherical wave calculations on crystalline $Li_{12}Si_7$ and provided new insights into the compound [83]. Their results basically match the interpretation of von Schnering et al., with the major difference that the five-atom ring is assigned a charge of −6 and the two remaining electrons fill the 2 s shell of the ring coordinating lithium atoms. This yields a formula of $(Li^+)_{22}(Li^-)_2(Si_4^{8-})(Si_5^{6-})_2$. Anyway, in this compound the "star" acts as accompanying anion which supports the N_E value where rings are stable (see Sect. 2.3).

The same three jagged star anion is reported in $Ba_5Mg_{18}E_{13}$ (N_E = 7.54 e/atom; $r(M_{av}^{I/II})$ = 0.092 nm) and $Sr_{6.33}Mg_{16.67}E_{13}$ (N_E = 7.54 e/atom; $r(M^{I/II})_{av}$ = 0.091 nm) for E = Si or Ge [84]. The N^E value for these compounds lies quite far away from the other reported values for compounds containing isolated polyanions. The crystal structures of the ternary materials show as additional anionic moieties the presence of monoatomic E^{4-} anions, and a composition of E_4:E^{4-} of 1:9 reveals the charge of the four-atom star to be −10. This gives rise to internal redox capacity of such E_4 groups. Beside mixed cation sites in the second compound, which lead to geometrically unusual positions resulting in very short Sr-E distances of 292 pm (Si) and 293 pm (Ge), the central atoms of the stars feature a very distinctive anisotropic displacement. For germanium, a slightly pyramidal Ge_4 group could be resolved, which is reflected in the electron density distributions. They show two distinct density maxima for the germanium position (Fig. 4a). Corresponding results could not be obtained for Si_4 from X-ray diffraction data, and ordering of the Si_4 group at lower temperatures could also be excluded. Therefore, a static disorder of Si_4 groups is assumed. Consequently, E_4 groups may occur as completely planar E_4^{8-} ions, isostructural and isoelectronic to CO_3^{2-}, or as slightly pyramidal E_4^{10-} ions, depending on the nature of the cations [84].

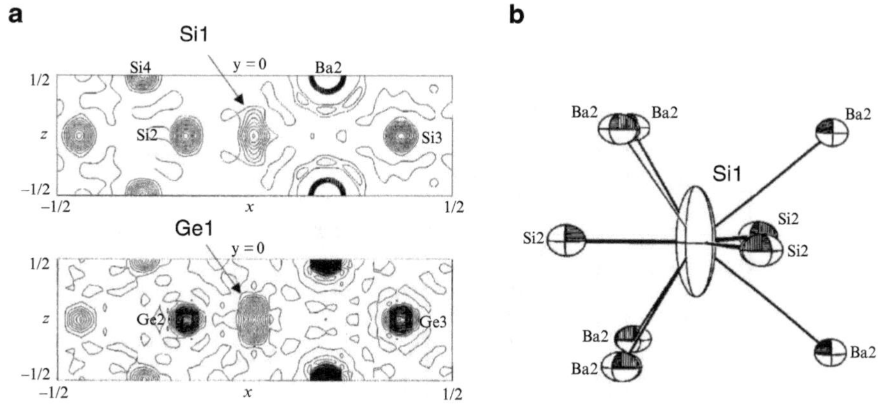

Fig. 4 (**a**) Charge density distribution and (**b**) coordination spheres of the star-like E_4 units in $Ba_5Mg_{18}E_{13}$ (E = Si, Ge). ([84] Copyright Wiley-VCH Verlag GmbH & Co KGaA. Reproduced with permission)

2.6 *Combination of Different Tetrel Elements*

Very recently, a mixture of silicon and tin reduced by barium metal yielded the compound Ba_8Si_6Sn, which contains three-atom rings of silicon atoms Si_3^{6-} and monoatomic Sn^{4-} [85]. This shows that the combination of differently sized anionic atoms gives the opportunity to expand into hitherto unknown (N_E; r(M)) regions that contain new polyanions.

2.7 *Summarized: Homoatomic Polyanions of Group 14 Elements*

To summarize the previously presented anions, the model of the averaged radii of cations in relation to the degree of reduction is applied, which results in areas of stability for the differently natured anions. Concerning clusters, large nine- and eight-atom clusters tend to be stabilized by rather large cations, which is in accordance with simple packing considerations as well as the HSAB principle. In contrast, the smaller tetrahedral E_4^{4-} clusters may be stabilized by a very broad range of cation radii. Compounds that contain tetrahedral anions beside the nonatetrelide cages show an expanded cation range compared to the pure nine-atom compounds. In ring-containing compounds, both the average radius of the cationic moieties and the degree of reduction generally are within a smaller array compared to the clusters. Five-membered rings of silicon and germanium are present within 0.06 and 0.08 nm cation size and a degree of reduction between 5.6 and 5.8 e/atom. Around 5.7 e/atom and 0.09–0.11 nm cation radius, the heavier

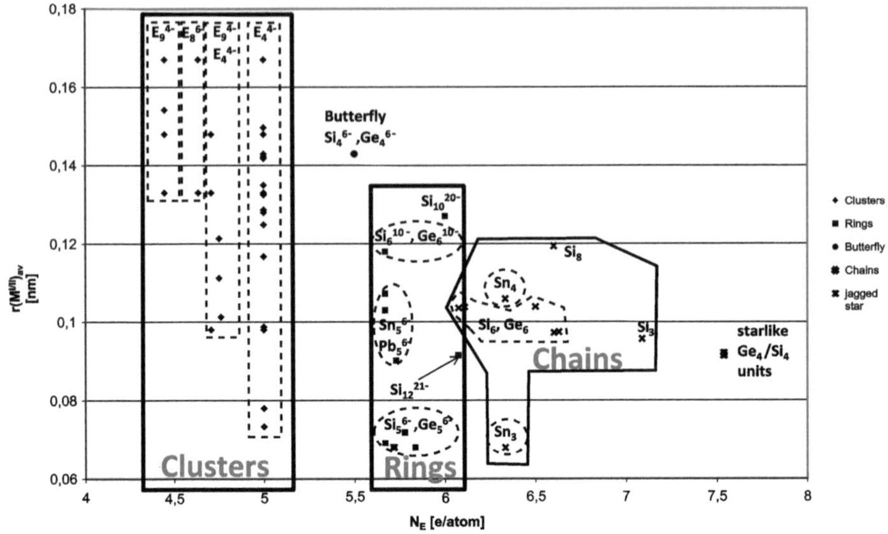

Fig. 5 Polyanions of group 14 and their dependence on averaged cation radii and degree of reduction

homologue rings of tin and lead are observed. An increased cation radius at a similar degree of reduction yields six-membered rings of silicon and germanium. At a higher degree of reduction, substituted derivatives of the six-atom rings of silicon are obtained. The combination of differently sized anion atoms provides the opportunity to push the limits further to compounds with higher N_E values which contain rings.

Concerning compounds with a still higher charge for the formally anionic structure part, between a degree of reduction of six and seven e/atom chains are present, where four to seven atom chains are stabilized best by an averaged cation radius around 0.1 pm. Three-atom chains are observed either at smaller cation radii or at a higher degree of reduction. Larger cations yield an eight-atom silicon chain. At a very high degree of reduction, the presence of a four-atom star is reported (Fig. 5).

3 Homoatomic Polyanions of Group 15 Elements in Solid State Compounds

Polyphosphide compounds within the scope of this overview rely heavily on contributions by von Schnering et al. (for a review on polyphosphides, see [86]), and major results concerning the heavier homologues As, Sb, and Bi are due to Röhr et al., Eisenmann et al., and Belin et al.

3.1 *Clusters of Group 15 Elements in Compounds with 5.2 < N_E < 5.5*

When combining electropositive alkali and alkaline earth metals with group 15 elements, the lowest degree of reduction for molecular polyanions is between 5.2 and 5.5 e/atom. This gives an average of 2.8 to 2.5 E–E contacts according to the 8-N rule, and is therefore less than in elemental structures, where each group 15 atom is threefold bound. For these N_E values, a combination between twofold and threefold bound atoms is expected in crystal structures, which in case of isolated species results in two different homoatomic cluster types. Table 6 contains homoatomic clusters in solid state compounds of group 15 elements.

In contrast to group 14 element clusters, here only two different cluster types are observed. The most common cluster species is represented by the nortricyclane-like E_7^{3-} anions (E = P–Sb), which were structurally characterized for the first time by von Schnering et al. in the compound Sr_3P_{14} in 1972 [94]. The N_E value in these materials adds up to 5.43 e/atom. The existence of the yellow binary alkali metal compounds M_3P_7 with isoelectronic anions was known for a long time [103–105], but crystal structures for Li_3P_7 [90] and Cs_3P_7 [93] were reported only in the middle of the 1980s, which is due to the crystalline to plastic–crystalline transitions of M_3P_7 materials (M = Na–Cs) [91]. Crystal structures of the analog compounds for the heavier homologue element arsenic even have been reported as late as 2002 [96]. The nortricyclane-like E_7^{3-} anions feature three main distances (A, B, C), and for

Table 6 Homoatomic clusters of group 15 elements in solid state compounds

		Compound	N_E	$r(M^{I/II})_{av}$	References
E_{11}^{3-}	P	Na_3P_{11}	5.27	0.098	[87]
		K_3P_{11}	5.27	0.133	[88]
		Rb_3P_{11}	5.27	0.148	[88]
		Cs_3P_{11}	5.27	0.167	[88]
	As	K_3As_{11}	5.27	0.133	[88]
		Rb_3As_{11}	5.27	0.148	[88, 89]
		Cs_3As_{11}	5.27	0.167	[88, 89]
E_7^{3-}	P	Li_3P_7	5.43	0.068	[90]
		Na_3P_7	5.43	0.098	[91]
		K_3P_7	5.43	0.133	[91]
		Rb_3P_7	5.43	0.148	[91, 92]
		Cs_3P_7	5.43	0.167	[93]
		Sr_3P_{14}	5.43	0.127	[94]
		Ba_3P_{14}	5.43	0.143	[95]
	As	Li_3As_7	5.43	0.068	[96, 97]
		Na_3As_7	5.43	0.098	[96, 98]
		K_3As_7	5.43	0.133	[96]
		Rb_3As_7	5.43	0.148	[96, 99]
		Cs_3As_7	5.43	0.167	[96]
		Ba_3As_{14}	5.43	0.143	[100]
	Sb	Rb_3Sb_7	5.43	0.148	[101]
		Cs_3Sb_7	5.43	0.167	[101, 102]

Fig. 6 E_7^{3-} and E_{11}^{3-} anions with the labels of the bond lengths, pertinent bond angles, and the heights of the polyanions

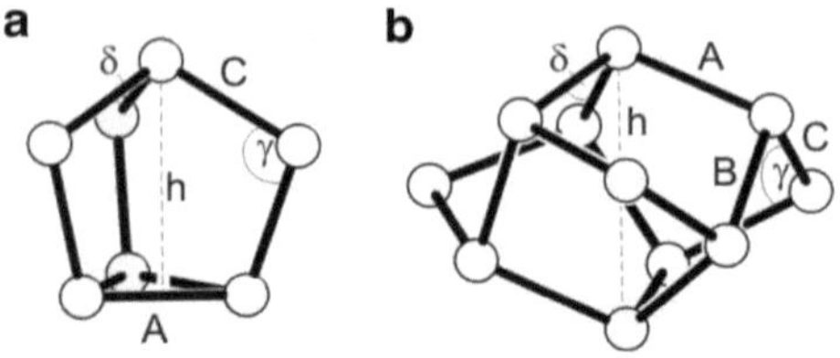

neutral heptaheteronortricyclanes P_4S_3 and P_4Se_3 $A > B > C$ and $\gamma > \delta$ holds true, whereas for the homoatomic ionic cages here discussed $A > C > D$ and $\delta > \gamma$ applies (Fig. 6a) [86].

The least-reduced isolated polyanion in a dense solid without alkali metal ligands or solvent molecules is represented by the E_{11}^{3-} anions (E = P, As), which feature a N_E value of 5.27 e/atom. For these kinds of trishomocubane-like clusters, a similar but less-pronounced analysis of bonds lengths and angles is applicable (Fig. 6b). The phosphorus compounds and the heavy alkali metal arsenides Rb_3As_{11} and Cs_3As_{11} can be obtained directly from the elements. For the preparation of the potassium analog compound, K_3As_7 is used as reducing agent for stoichiometric amounts of arsenic.

3.2 *Rings of Group 15 Elements in Compounds with $N_E = 5.67, 6$*

An increased N_E value results in the formation of isolated rings of group 15 elements in solid state structures. Besides these rings, of course two-dimensionally extended layers and one-dimensionally extended chains are observed, according to the 8-N rule for these N_E values. In Table 7, solid state compounds are listed in which homoatomic rings are present in the crystal structure.

There are three kinds of rings of different nuclearity known for homoatomic group 15 anions (Fig. 7). For arsenic, three-, five-, and six-atom rings are described in solid state compounds, and for phosphorus only the six-atom ring is reported, whereas for antimony and bismuth no homoatomic rings are known in neat solid state materials of alkali and alkaline earth metals.

Rings of six atoms are observed for phosphorus and arsenic as the exclusive anionic moiety in compounds following the formula $M_4^IE_6$. This implies a charge of −4 for the rings, and chemical bonding referring to the Hückel concept was carefully discussed due to planarity of the anion in its crystal structures [107, 111] (Fig. 8).

This planarity is caused by the crystal symmetry of the space group *Fmmm* (*Fddd* for β-K_4P_6). Recently, quantum chemical calculations could demonstrate for P_6^{4-} that the ring is slightly distorted in a chair-like conformation, which is

Table 7 Homoatomic rings of group 15 elements in solid state structures

	Compound	N_E	$r(M^{I/II})_{av}$	References
P_6^{4-}	K_4P_6	5.67	0.133	[106]
	Rb_4P_6	5.67	0.148	[107, 108]
	Cs_4P_6	5.67	0.167	[107]
As_6^{4-}	Rb_4As_6	5.67	0.148	[109]
	Cs_4As_6	5.67	0.167	[109]
As_3^{3-}	CsAs	6	0.167	[96]
As_5^{5-}	Ba_2KAs_5	6	0.140	[110]
	Ba_2RbAs_5	6	0.145	[110]

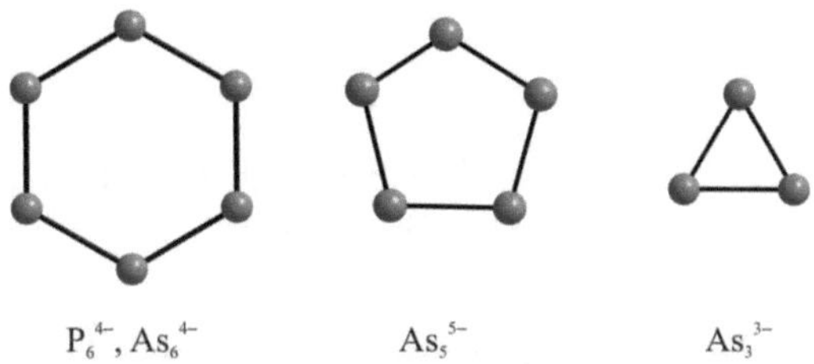

Fig. 7 Rings of group 15 elements in solid state compounds of alkali and alkaline earth metals

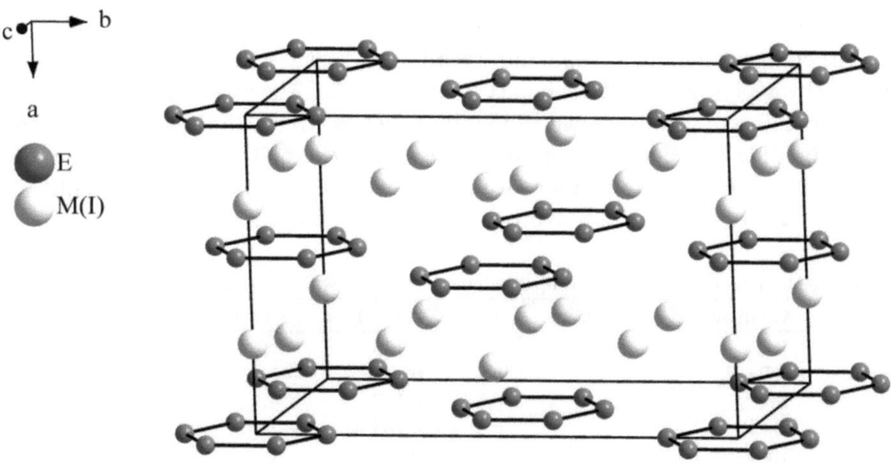

Fig. 8 Unit cell of $M_4^IE_6$ (M^I = K–Cs; E = P, As) contains nonaromatic six atom rings

contradictory to the previously assumed aromatic character but fits perfectly to the also missing low field shift of the anions in ^{31}P MAS NMR spectra [112, 113].

With arsenic, five-atom rings in an envelope conformation are observed in compounds with a N_E value of 6, which consequently carry a charge of −5 corresponding to twofold bound atoms. These rings only are present in the mixed alkali/alkaline earth metal ternary materials $Ba_2M^IAs_5$ (M^I = K, Rb). Only one binary alkali metal pentelide phase containing isolated anions is reported at the same N_E value, which contains three-atom rings of arsenic together with caesium.

For the compounds with lighter alkali metals which have the same composition, the formation of the isoelectronic one-dimensionally extended chains is favored [96]. This once again demonstrates the effect of cation size beyond the charge compensating function on the nature of the observed anions.

By taking the cation radii into account, it becomes evident that rather large cationic moieties are necessary to stabilize isolated rings of group 15 elements in solid state compounds. Smaller cations lead to extended structures as their size does not suffice to isolate discrete anions. Therefore, rings are only observed in compounds of cations with radii above 0.133 nm for phosphorus and above 0.140 nm for the heavier homologue arsenic. In contrast, group 14 rings are observed up to an averaged cation radius of 0.127 nm.

3.3 *Chains of Group 15 Elements in Compounds with 6.11 ≤ N_E ≤ 6.8*

While the number of differently shaped clusters and rings of group 15 elements is quite limited, compounds with N_E values between 6.11 and 6.8 e/atom display an impressive variety of dissimilar discrete chains. In Table 8, chain-containing solid state structures of group 15 elements are listed.

The charge assignment is straightforward for P_3^{5-}, E_4^{6-} (E = P, As), Sb_6^{8-}, and As_8^{10-} by applying the valence bond concept (Fig. 9a). The bonding situation of the electron-deficient chains shown in Fig. 9b is not as trivial. In these cases, on one hand, multiple element–element bonding is discussed, and on the other hand, the possibility of partial protonation cannot be excluded completely, especially for compounds of the sometimes hydrogen-contaminated metal barium.

Table 8 Homoatomic chains of group 15 elements in solid state structures

Chain	Compound	Add. anion	N_E	$r(M^{I/II})_{av}$	Electron precise[a]	References
P_3^{4-}	K_4P_3	–	6.33	0.133	–	[114]
P_3^{5-}	KBa_4P_5	P_2^{4-}	6.8	0.141	+	[115]
P_4^{6-}	Sr_3P_4		6.5	0.127	+	[116]
	Ba_3P_4	–	6.5	0.143	+	[116]
P_9^{10-}	Ba_5P_9	–	6.11	0.143	–	[117]
As_4^{5-}	K_5As_4	–	6.25	0.133	–	[118]
As_4^{6-}	Ca_2As_3	As_8^{10-}	6.33	0.106	+	[119]
	Sr_3As_4	–	6.5	0.127	+	[120]
As_8^{10-}	Ca_2As_3	As_4^{6-}	6.33	0.106	+	[119]
Sb_4^{5-}	K_5Sb_4	–	6.25	0.133	–	[121]
	Rb_5Sb_4	–	6.25	0.148	–	[118]
Sb_6^{8-}	Ba_2Sb_3	–	6.33	0.143	+	[122]
	Sr_2Sb_3	–	6.33	0.127	+	[122]
Bi_4^{5-}	K_5Bi_4	–	6.25	0.133	–	[118]
	Rb_5Bi_4	–	6.25	0.148	–	[118]
	Cs_5Bi_4	–	6.25	0.167	–	[118]

[a]Electron count according to the (8-N) rule and the *Zintl–Klemm* concept

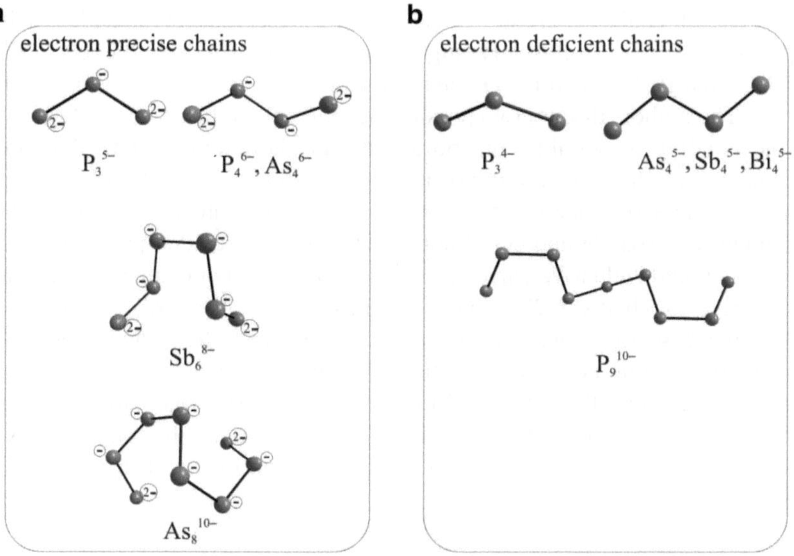

Fig. 9 (**a**) Electron precise chains of group 15 elements in solid state structures with formal charges assigned by valence count; (**b**) electron-deficient chains of group 15 elements; all chains are represented in perspective view to convey the helical assemblies

For three- and four-atom chains, electron precise and electron-deficient moieties are known. Three-atom chains are only observed in phosphorus-containing materials. The existence of an electron precise P_3^{5-} chain is reported in the mixed alkali/alkaline earth metal material KBa_4P_5. The structure exhibits a 50/50 anionic occupational disorder in a single cavity between the cations, and as an additional anion in this cavity, P_2^{4-} dumbbells are described [115]. The second *catena*-triphosphide is present in K_4P_3, where the three-atom chain is assigned a fourfold negative charge, which implicates that the anion is a radical. The expected paramagnetism was proven. The FIR P–P valence vibrations are shifted 30–40 cm^{-1} toward higher wave numbers compared to other P–P valence vibrations. This is in agreement with the increased bond order [114].

Electron precise angulated four-atom chains are found in the isotypic phosphorus and arsenic alkaline earth metal compounds Sr_3P_4, Ba_3P_4, and Sr_3As_4. $M_3^{II}E_4$ was shown to be strongly related to the α-$ThSi_2$ structure type in terms of the formulation $M_3^{II}E_4\square_2$ ($\square$ represents voids in the anionic partial structure of a-$ThSi_2$) [116]. Additionally, As_4^{6-} chains are observed in Ca_2As_3 next to As_8^{10-} chains. For the heavier group 15 elements As, Sb, and Bi, a one-electron oxidized variant of the four-atom chain is known. As the chains are flat, a description by single bonds seems implausible. The two possibilities that were discussed are the delocalization of one electron over the isolated anions or over the whole structure. To elucidate the electronic structure, magnetic measurements were carried out, which resulted in the detection of temperature-independent *Pauli*-like paramagnetism. This puts the compounds in the rare class of *metallic salts* [118].

The longest chain of group 15 elements is observed for phosphorus in the compound Ba_5P_9. For exclusively single-bonded atoms, one would expect a charge of -11; so the P_9^{10-} chain is one electron short for a description according to the 8-N rule. The authors discuss two possibilities to explain this feature. On one hand, multiple bonding is possible, but no significant shortening of one P–P bond is observed, and the helical structure of the anion also is contrary to this description. On the other hand, protonation of one phosphorus atom in an electronically balanced P_9^{11-} might be possible and could not be completely excluded [117]. The second description would yield a N_E value of 6.22 e/atom which would nicely fit and narrow the N_E region for chains ($6.22 \le N_E \le 6.8$ instead of $6.11 \le N_E \le 6.8$).

As is the case for group 14 chains, the N_E values where chain formation is observed span a quite large range. A first comparison between group 14 and group 15 chains concerning the cation radii clearly shows that large cation radii above 0.106 nm are needed to stabilize group 15 element chains. In contrast, group 14 element chains are observed below an averaged cation radius of 0.106 nm.

3.4 A Branched P_8^{10-} Anion at $N_E = 6.25$

Within the N_E region of isolated chain fragments, an additional phosphide anion was observed by Hadenfeldt and Bartels [123]. In Ca_5P_8, the combination with the small calcium cations ($r(Ca^{2+})$= 0.106 nm) results in a polyphosphide that contains fourfold-bound phosphorus atoms beside single-bonded atoms. The centrosymmetric anion P_8^{10-} possesses 50 valence electrons, and it is shaped like ethane in the staggered conformation. This is the first example of a polyphosphide which contains formally positively charged phosphorus atoms beside formally twofold negatively charged (Fig. 10). P_8^{10-} is isoelectronic to the hexathiophosphate anion $P_2S_6^{4-}$ and isovalence electronic to staggered $Si_2Te_6^{6-}$ as well as staggered or eclipsed $Ga_2X_6^{2-}$ (X = Cl, Br, I) anions [124].

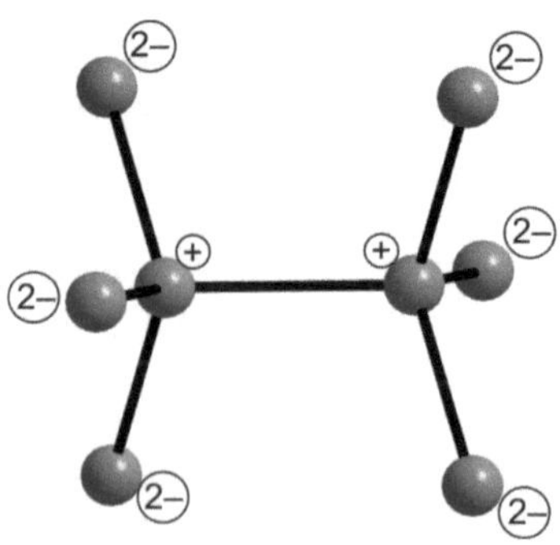

Fig. 10 P_8^{10-} anion in Ca_5P_8 (with assigned formal charges) features an ethane-like staggered confirmation

3.5 *Summarized: Homoatomic Polyanions of Group 15 Elements*

The model of the averaged radii of cations in relation to the degree of reduction is applied, which results in areas of stability for the differently natured anions of group 15, similar to the observations for the group 14 element anions.

Homoatomic group 15 clusters are observed at $N_E = 5.27$ (E_{11}^{3-}) and 5.43 (E_7^{3-}). The larger E_{11}^{3-} clusters tend to be stabilized by larger cations; only for the smaller phosphorus cage E_7^{3-} a crystal structure of a sodium-containing compound is reported, which again is in accordance with simple packing considerations as well as the HSAB principle. Other E_{11} cage compounds (E = P, As) are only stabilized from a cation radius of 0.133 nm (potassium) upward. No Sb_{11}^{3-} cluster has been reported in neat solid state compounds. In contrast to this, smaller nortricyclane-like E_7^{3-} clusters are observed for cation radii between 0.068 nm (Li) and 0.167 nm (Cs).

At a N_E value of 5.67 e/atom, six-membered rings E_6^{4-} are present in crystal structures for cation radii between 0.133 nm (K) and 0.167 nm (Cs). Further reduced 1:1 compounds of phosphorus and arsenic for alkali metals yield three- and five-membered rings for large cations. For larger N_E values from 6.11 e/atom (respectively 6.22 e/atom, see Sect. 3.3) on, the field of chain fragments affiliates, followed by phosphorus assemblies containing nine or ten atoms which are present around 6.25 e/atom. For larger cations and heavier group 15 elements, electron-deficient E_4^{5-} chains are observed in binary compounds. Further reduction yields for phosphorus the electron-deficient P_4^{3-} anion, and for arsenic and antimony

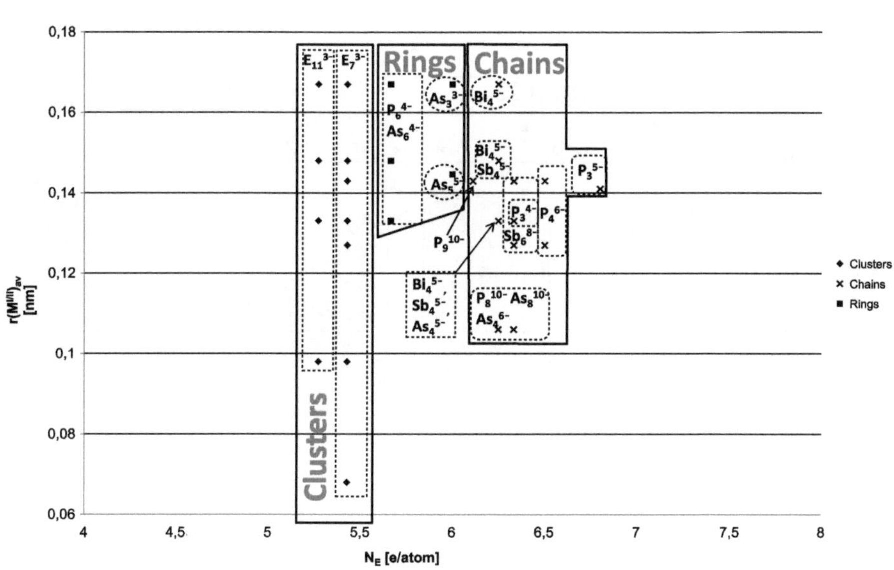

Fig. 11 Homoatomic polyanions of group 15 elements in neat solids

larger electron precise chain fragments of four, six, or eight atoms. At $N_E = 6.5$ e/atom, electron precise E_4^{6-} chains for phosphorus and arsenic are observed. After a large gap in N_E value, KBa_4P_5 is reported, which includes occupational disordered P_2^{4-} dumbbells and P_3^{5-} chains on the same crystallographic site.

4 Combining the Jigsaw Puzzle: Homoatomic Polyanions of Group 14 and Group 15 Elements in Solid State Compounds

We introduced (N_E; $r(M^{I/II})_{av}$) pairs to organize known solid state compounds which include homoatomic polyanions of group 14 and group 15 elements. The schemata given in Figs. 5 and 11 show regions that designate cluster, ring, and chain formation. Of course, these structures always compete with extended isoelectronic networks. Combining now both diagrams of group 14 and group 15 persuasively shows how homoatomic polyanions of the one group complement and accompany homoatomic polyanions of the other group. Clusters of group 15 elements follow the cluster compounds of group 14, as one would expect due to the 8-N rule. In contrast to that, rings and chains are present for both groups of the periodic table at similar N_E values. This gives a narrow N_E value for rings between 5.6 and 6.2 e/atom for elements of both groups. Subsequently, the formation of chains is observed up to high degrees of reduction of 7.5 e/atom, if one counts

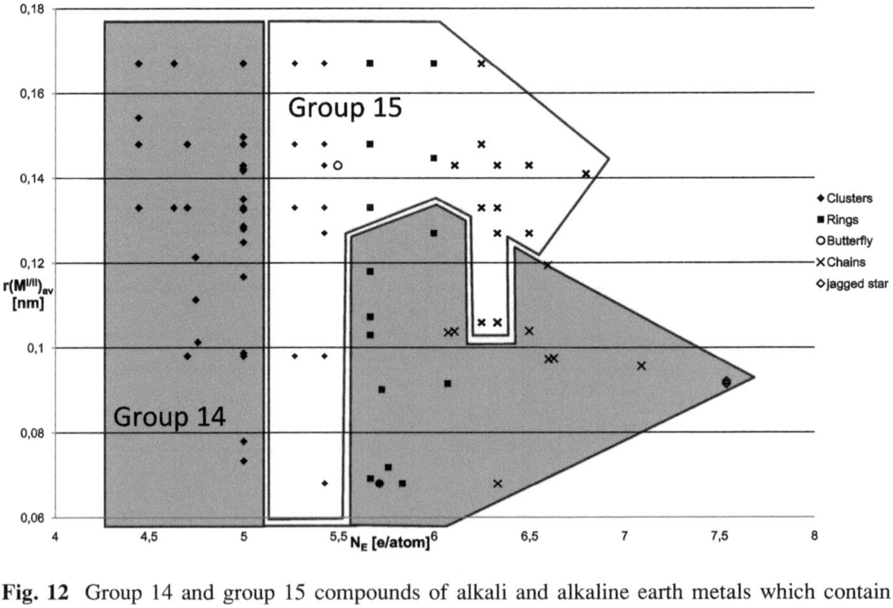

Fig. 12 Group 14 and group 15 compounds of alkali and alkaline earth metals which contain isolated homoatomic polyanions harmonize like parts of a jigsaw puzzle

branched chains as well. The majority of chains are present at N_E values between 6.2 and 6.7 e/atom. In contrast to this similarity, group 14 and group 15 ring- and chain-containing compounds differ significantly in the size of the involved cation radii. This is a remarkable characteristic of the applied structure mapping procedure. With respect to the known polyanionic compounds, no overlap is observable in (N_E; $r(M^{I/II})_{av}$) regions: Group 14 rings and chains crystallize together with small cations, whereas group 15 chains and rings seem to be rather stabilized by larger cations. Both the polyanionic regions of group 14 and group 15 can be seen as part of a jigsaw puzzle, which fits remarkably well and demonstrates the similarities and the differences of neighbors in the periodic table (Fig. 12).

5 Group 14 and Group 15 Polyanions in Solution

The presence of isolated anionic species coordinated by electropositive metal cations suggests the possibility to transfer both congruently in solutions using an appropriate solvent. This transfer is the ultimate test as to whether a polyanion perceived in a solid is really a separate and viable chemical entity, and of course the prerequisite for chemical transformations. The overview on binary and multinary polyanionic compounds given in the Chaps. 1–4 thus has to be complemented by an overview of crystalline solids containing solvent molecules and alkali metal ligands which were obtained from solutions, and which give a good indication of the species available for further chemistry.

As the first experiments on polyanionic compounds used liquid ammonia as the reaction medium, using this solvent suggests itself. The application of the ammonia-related solvent ethylenediamine provides the possibility to work at moderate temperatures. Up to now, a quite impressing number of solvate structures has been reported, which contain homoatomic polyanions and demonstrate the stability of these discrete element building blocks outside of a dense cation–anion network. Besides stable polyanions in solution, it has to be noted that due to their high degree of reduction, most higher charged anions are very sensitive toward oxidation. This oxidation may be accompanied by a rearrangement of the polyanions, and therefore some polyanions from the solid state are not observed in solution. In contrast, new polyanions that are not known from solid state compounds may be crystallized from solution and characterized by X-ray diffraction studies.

5.1 Group 14 Polyanions in Solution

In the past decades, homoatomic polyanions in solid state compounds of group 14 elements have been a powerful starting material toward different kinds of chemical transformations. Recent developments, especially for nine-atom clusters, have been discussed elsewhere [5, 125, 126] and are not within the scope of this review. Here, we concentrate on "neat solvates", which means that in addition to cation and anion

Table 9 Reported neat solvate structures of polyanions of group 14 elements, which are existent in solid state as well as in solution

Anion	Compound	N_E	References
Si_9^{4-}	$Rb_4Si_9 \cdot 4.75NH_3$	4.44	[127]
	$Rb_4Si_9 \cdot 5NH_3$	4.44	[128]
Ge_9^{4-}	$K_4Ge_9 \cdot 9NH_3$	4.44	[129]
	$Rb_4Ge_9 \cdot 5NH_3$	4.44	[129]
	$Rb_4Ge_9 \cdot en^a$	4.44	[130]
	$Cs_4Ge_9 \cdot en$	4.44	[131]
Sn_9^{4-}	$Na_4Sn_9 \cdot 7en$	4.44	[132, 133]
	$[Li(NH_3)_4]_4Sn_9 \cdot NH_3$	4.44	[134]
Pb_9^{4-}	$[Li(NH_3)_4]_4Pb_9 \cdot NH_3$	4.44	[134]
Sn_4^{4-}	$Rb_4Sn_4 \cdot 2NH_3$	5	[135]
	$Cs_4Sn_4 \cdot 2NH_3$	5	[135]
Pb_4^{4-}	$Rb_4Pb_4 \cdot 2NH_3$	5	[135]

Here, only nine-atom cages without sequestering agents like crown ethers or cryptands are listed; for an exhaustive overview on E_9 compounds, see [5, 125, 126]

[a]This compound was obtained by exchange reaction of a solution of "$NaGe_{2.25}$" (precursor) in ethylenediamine with solid RbI at 363 K

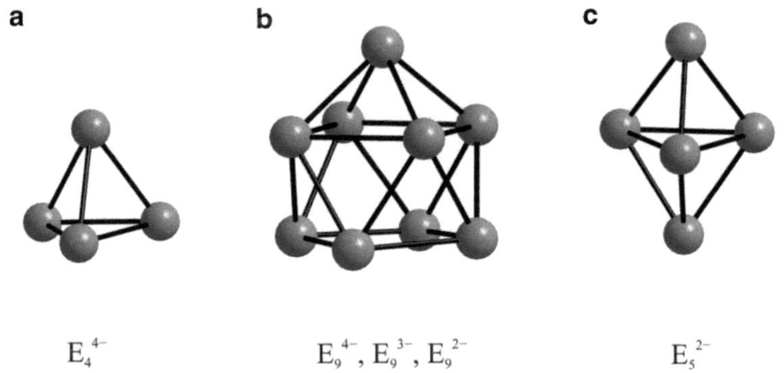

Fig. 13 Homoatomic polyanions of group 14 elements, which are present in solvate structures; (**a**) tetrahedral E_4^{4-} (E = Sn, Pb); (**b**) nine-atom cages E_9^{n-} (n = 2–4; E = Si–Pb); (**c**) trigonal bipyramidal E_5^{2-} (E = Si–Pb)

from the polyanionic salts, exclusively solvate molecules participate in the crystal structures. Table 9 contains polyanions of group 14 elements, which are present in solvate structures and therefore presumably are stable not only in the solid state but also in the solution.

Having a closer look at the reported solvate structures of homoatomic polyanions of group 14 elements, it becomes evident that only clusters and no rings or chains are stable in solid state as well as in solution (Fig. 13). Higher reduced polyanions with $N_E > 5$, equivalent with a charge higher than -1 per atom, are not reported in solution so far. The nine-atom species is quite well established in solutions of ammonia and ethylenediamine. Their interesting reactivity is intensely studied and is discussed elsewhere [5, 125, 126]. The less-understood cluster species in solution is represented by the four-atom tetrahedral anions, which are only present in solutions of ammonia for the elements tin and lead. Recently, it was proposed that the four-atom species are the origin of the trigonal bipyramidal five-atom clusters [128]. This cluster species E_5^{2-} (E = Si [128, 136]; Ge [137, 138]; Sn [139, 140]; Pb [140, 141]) is only known in solvate structures in combination with large alkali metal cryptand complexes in compounds following the compositions $[M^I@crypt]_2E_5$ (E = Sn, Pb) and $[M^I@crypt]_2E_5 \cdot 4NH_3$ (E = Si, Ge). Nonatetrelide cages are known in different oxidation states (E_9^{2-}, E_9^{3-}, E_9^{4-}), where the less-charged cages E_9^{2-} and E_9^{3-} tend to crystallize together with sequestering agents like for example 18-crown-6 or [2.2.2]-cryptand. In all compounds, the number of atoms remains the same, independently from charge, and merely the shapes of the clusters change. Therefore, the nine-atom species represents the most stable cluster species in polyanionic salt solutions, which can compensate oxidation and reduction processes without changing the cluster size. This is also corroborated by ^{119}Sn solution NMR studies, where the signal of the nine-atom clusters at −1,224 ppm increases after several days [142, 143]. In contrast, the four-atom cages compensate oxidation processes by rearrangement and enlargement of the cluster.

5.2 Group 15 Polyanions in Solution

In solvate structures of polypentelides, some polyanions are observed which are also present in solid state structures. These polyanions are in most instances limited to the 7- and 11-atom cluster compounds (Table 10). Using ion exchange resins, alkali metal cations can be replaced by quaternary ammonium or phosphonium cations. The shape of the clusters in all compounds is very similar to the ones found in the solid state materials. Only one homoatomic ring which is also known in solid state materials has been found up to now, and it is present in the compound $[Rb(18\text{-crown-}6)]_2Rb_2As_6 \cdot 6NH_3$. In contrast to the solid state compound Rb_4As_6, where the rings are planar due to space group symmetry, in the solvate structure the rings are slightly distorted. This distortion agrees well with quantum chemical calculations which confirmed the absence of aromaticity in these rings [112].

In addition to the recrystallization of polyanionic salts from solution, the dissolution and oxidation of these compounds provide the possibility of generating new polyanions, which consequently are only accessible by crystallization from solution. The hitherto known compounds that contain new anionic species are listed in Table 11. The formation of new ring anions is observed. Six-membered ring containing A_4P_6 materials dissolve incongruently in liquid ammonia under the formation of heptaphosphide cages and lone-pair aromatic P_4^{2-} rings [112, 145, 169]. Other rings of group 15 elements crystallizing from solution are a Sb_5^{5-} ring in envelope conformation in $[Li(NH_3)_4]_3[Li_2(NH_3)_2Sb_5] \cdot 2NH_3$ and a S_8-like Sb_8^{8-}anion in $[K_{17}(Sb_8)_2(NH_2)] \cdot 17.5NH_3$. Additionally, cyclic As_4^{2-} and Sb_4^{2-} anions are also reported (Fig. 14).

Table 10 Polyanions of group 15 in solvate structures, which are also known in solid state compounds

Anion	Compound	N_E	References
P_7^{3-}	$Cs_3P_7 \cdot 3NH_3$	5.43	[144]
	$Rb_3P_7 \cdot 7NH_3$	5.43	[145]
	$Ba_3P_{14} \cdot 18NH_3$	5.43	[146]
	$[NEt_3Me][Cs_2P_7] \cdot NH_3$	5.43	[147]
	$[NEt_4]Cs_2P_7 \cdot 4NH_3$	5.43	[147]
	$[NEtMe_3]Cs_2P_7 \cdot 2NH_3$	5.43	[148]
	$[NMe_4]_2RbP_7 \cdot NH_3$	5.43	[149]
	$[Rb(18\text{-crown-}6)]_3P_7 \cdot 6NH_3$	5.43	[145]
	$[K(18\text{-crown-}6)]_3K_3(P_7)_2 \cdot 10NH_3$	5.43	[145]
	$Li_3P_7 \cdot 3TMEDA$	5.43	[150]
P_{11}^{3-}	$Cs_3P_{11} \cdot 3NH_3$	5.27	[151]
	$BaCsP_{11} \cdot 11NH_3$	5.27	[152]
	$[NEt_3Me]_2CsP_{11} \cdot 5NH_3$	5.27	[153]
	$[NEtMe_3]_3P_{11}$	5.27	[154]
	$[NEt_4]Cs_2P_{11}$	5.27	[149]
	$[K(18\text{-crown-}6)]_3P_{11} \cdot 2en$	5.27	[155]
As_6^{4-}	$[Rb(18\text{-crown-}6)]_2Rb_2As_6 \cdot 6NH_3$	5.67	[112]
As_7^{3-}	$[Li(NH_3)_4]_3As_7 \cdot NH_3$	5.43	[156]
	$Cs_3As_7 \cdot 6NH_3$	5.43	[156]
	$Cs_3As_7 \cdot NH_3$	5.43	[157]
	$[Li(TMEDA)]_3As_7 \cdot OEt_2$	5.43	[158]
	$[Li(TMEDA)]_3As_7 \cdot 1.5\ tol$	5.43	[159]
	$[Li(DME)]_3As_7 \cdot OEt_2$	5.43	[160]
	$[NMe_4]_2RbAs_7 \cdot NH_3$	5.43	[161]
	$[Rb(18\text{-crown-}6)]_3As_7 \cdot 8NH_3$	5.43	[156]
	$[P(Ph_3)_4]_2CsAs_7 \cdot 5NH_3$	5.43	[156]
	$[K([2.2.2]\text{-crypt})]_{1.5}K_{1.5}As_7$	5.43	[162]
As_{11}^{3-}	$[Cs(18\text{-crown-}6)]_2CsAs_{11} \cdot 8NH_3$	5.27	[163]
	$[K([2.2.2]\text{-crypt}]_3As_{11}$	5.27	[164]
Sb_7^{3-}	$[Na([2.2.2]\text{-crypt})]_3Sb_7$	5.43	[165]
	$Na_3Sb_7 \cdot 3en$	5.43	[132]
	$Li_3Sb_7 \cdot 3TMEDA \cdot tol$	5.43	[166]
	$Na_3Sb_7 \cdot 3TMEDA \cdot 3thf$	5.43	[159]
	$Li_3Sb_7 \cdot 6HNMe_2$	5.43	[166]
	$[K([2.2.2]\text{-crypt})]_3Sb_7 \cdot 2en$	5.43	[167]
	$Na_3Sb_7 \cdot 3PMDETA \cdot tol$	5.43	[168]

TMEDA tetramethylethylenediamine, *PMDETA* pentamethyldiethylenediamine

Contrary to solid state compounds, where E_{11}^{3-} anions are only known for phosphorus and arsenic, in solution Sb_{11}^{3-} clusters are accessible, which show similar structural features as the lighter homologue As_{11}^{3-} [163]. Using oxidizing agents, P_7^{3-} and As_7^{3-} clusters can be connected by single bonds to form dimeric P_{14}^{4-} and As_{14}^{4-} anions. The same is possible for undecapentelides, and solvate structures of As_{22}^{4-} and P_{22}^{4-} have been reported (Fig. 15).

Protonation of polyanions of group 14 and group 15 elements generally means an oxidation of the latter, due to the higher electronegativity of hydrogen compared to the heavier main group elements. For phosphorus and arsenic, the electronegativities are very similar to the value for hydrogen; therefore, one can expect stable

Table 11 Polyanions that are (up to now) not known in solid state compounds and therefore are only accessible from solution

Anion	Compound	N_E	References
P_4^{2-}	$Cs_2P_4 \cdot 2NH_3$	5.5	[145, 169, 170]
	$(K@18\text{-Krone-6})_2P_4 \cdot 8.5NH_3$	5.5	[145, 169]
P_{14}^{4-}	$[Li(NH_3)_4]_4P_{14} \cdot NH_3$	5.29	[171]
P_{22}^{4-}	$(NEtMe_3)_4P_{22} \cdot 2NH_3$	5.18	[172]
As_4^{2-}	$[Li(NH_3)_4]_2As_4$	5.5	[173]
	$[Na(NH_3)_5]_2As_4 \cdot 3NH_3$	5.5	[173]
	$[Cs_{0.35}Rb_{0.65}(2,2,2\text{-crypt})]_2As_4 \cdot 2NH_3$	5.5	[173]
As_{14}^{4-}	$[Rb(18\text{-crown-6})]_4As_{14} \cdot 6NH_3$	5.29	[171]
As_{22}^{4-}	$[Rb([2.2.2]\text{-crypt})]_4As_{22} \cdot 4DMF$	5.18	[174]
Sb_4^{2-}	$[K([2.2.2]\text{-crypt})]_2Sb_4$	5.5	[167]
Sb_5^{5-}	$[Li(NH_3)_4]_3[Li_2(NH_3)_2Sb_5] \cdot 2NH_3$	6	[175]
Sb_8^{8-}	$[K_{17}(Sb_8)_2(NH_2)] \cdot 17.5NH_3$	6	[176]
Sb_{11}^{3-}	$[K(18\text{-crown-6})(NH_3)_2]_3Sb_{11} \cdot 5.5NH_3$	5.27	[163]
	$[Na([2.2.2]\text{-crypt})]_3Sb_{11}$	5.27	[177]

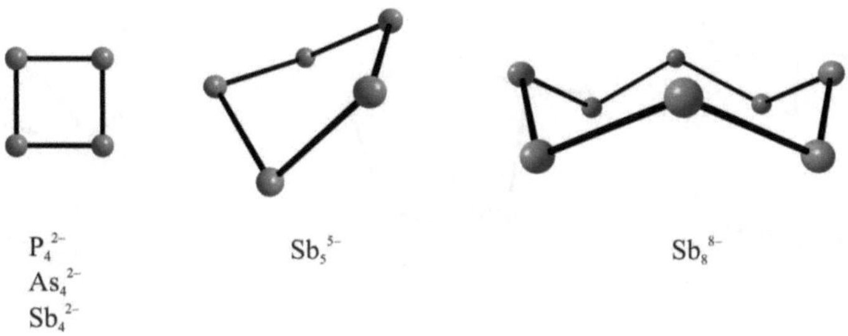

Fig. 14 Group 15 rings crystallized from ammonia or ethylenediamine solutions

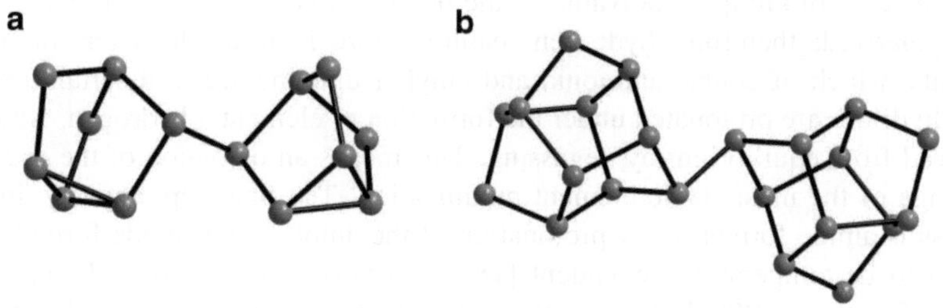

Fig. 15 Connected clusters for E = P, As: (**a**) E_{14}^{4-}, (**b**) E_{22}^{4-}

hydrogen polyphosphides and arsenides. The existence of hydrogen polypentelides in ammonia or ethylenediamine solution was proven and extensively studied by Baudler et al. using ^{31}P NMR techniques [178, 179]. Some of the predicted polyanions in solution have been crystallized and are summarized in Table 12 (Fig. 16).

Table 12 Hydrogenpentelides crystallized from solution

Anion	Compound	N_E	References
$P_3H_2^{3-}$	$K_3(P_3H_2) \cdot 2.3NH_3$	6	[180]
	$Rb_3(P_3H_2) \cdot NH_3$	6	[180]
$P_3H_3^{2-}$	$[Na(NH_3)_5][Na(NH_3)_3(P_3H_3)$	5.67	[181]
	$[Rb(18\text{-crown-}6)]_2(P_3H_3) \cdot 7.5NH_3$	5.67	[180]
	$[Cs(18\text{-crown-}6)]_2(P_3H_3) \cdot 7NH_3$	5.67	[180]
HP_7^{2-}	$[K([2.2.2]\text{-crypt})]_3K(HP_7)_2 \cdot en$	5.28	[182]
	$[K(18\text{-crown-}6)]_2HP_7$	5.28	[182]
	$[K(db18\text{-crown-}6)]_2HP_7 \cdot tol$	5.28	[182]
	$(PPh_4)_2HP_7 \cdot 3NH_3$	5.28	[183]
$H_2P_7^-$	$(PPh_4)H_2P_7$	5.14	[184]
HP_{11}^{2-}	$[K(2,2,2\text{-crypt})]_2HP_{11}$	5.18	[185]
	$(NBnMe_3)_2HP_{11}$	5.18	[154]
	$(PBnPh_3)_2HP_{11}$	5.18	[154]
	$[Sr(NH_3)_8]HP_{11} \cdot NH_3$	5.18	[186]

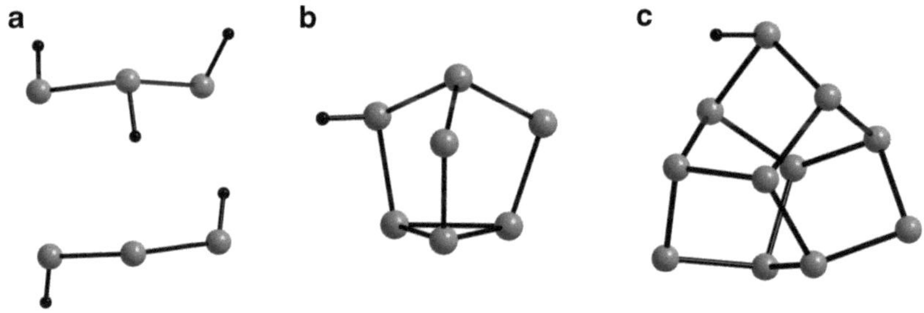

Fig. 16 Hydrogenpolyphosphides crystallized from ammonia solutions; (**a**) $P_3H_3^{2-}$, $P_3H_2^{3-}$; (**b**) HP_7^-; (**c**) HP_{11}^{2-}

In the case of group 14 polyanions, the bond between element and hydrogen is very polarized; therefore, hydrogen features more hydride character. In protic solvents, which of course ammonia and ethylenediamine are to a certain extent, these hydrides are protonated under the formation of elemental hydrogen, which is removed from equilibrium by degassing. This means an oxidation of the cages to the stage of the most stable element modification. The first step, anyway, in this process is amide formation by protonation of the anions. This amide formation is known to be temperature dependent [187] and very sensitive toward catalyzing traces of protons [188, 189]. Therefore, any source of protons has to be strictly avoided when working with homoatomic polyanions of group 14 in solution.

5.3 Summarized: Solvate Compounds of Group 14 and Group 15

If one compares the N_E values of solvate compounds that contain homoatomic polyanions, it becomes evident that there seems to be a border that has not been transcended so far. Group 14 compounds that have been crystallized from solution show a maximum N_E value of 5, corresponding to one negative charge per atom and threefold bound atoms. Group 15 element anions show a corresponding behavior by covering N_E values below 6, which also means a maximum value of one negative charge per group 15 atom and twofold bound atoms. Protonation of polyphosphides yields hydrogenpolyphosphide compounds, where the N_E value is further reduced compared to the homoatomic anions. For group 14 compounds, no protonated species have been reported so far. Here, different charges in E_9 clusters are compensated without any change in composition, whereas the four-atom tetrahedral clusters lead to larger cluster sizes when oxidized. When working with bare group 14 polyanions in solution, smallest traces of protons have to be avoided due to protonation being equivalent with oxidation of the anions. Smooth oxidation using, e.g., $SbPh_3$ or PPh_3 as catalysts provides the possibility of yielding new polyanionic materials.

6 Conclusions and Outlook

Main group elements in negative oxidation states are very interesting materials, which show all different kinds of anionic arrangements. In this review, we focused on discrete homoatomic polyanions, which have been ordered with respect to the degree of reduction and the averaged cation radii. This arrangement displays certain regions of stability of differently natured polyanions, which of course always compete with isoelectronic extended anionic subunits. It was shown that how regions of group 14 and group 15 polyanions fit like parts of a jigsaw puzzle in a combined diagram.

Polyanionic compounds in solution are a very fascinating and versatile class of compounds, the behavior of which is not understood completely. It is very challenging to tune the degree of oxidation for obtaining bigger discrete polyanionic units. On the other side, the limit for stable highly charged element building blocks in solution needs to be pushed further toward higher reduced species. It will be interesting to see whether the border defined by a maximum of one negative charge per atom can be crossed in the future for dissolved polyanions.

References

1. Joannis AC (1892) C R Acad Sci 114:587
2. Joannis AC (1891) C R Acad Sci 113:795
3. Zintl E, Dullenkopf W (1932) Z Phys Chem B16:183
4. Zintl E, Goubeau J, Dullenkopf W (1931) Z Phys Chem A154:1
5. Scharfe S, Fässler TF (2010) Philos Trans R Soc A 368:1265
6. Sevov SC (2002) Abstr Papers Am Chem Soc 224:614
7. Ugrinov A, Sevov SC (2002) J Am Chem Soc 124:10990

8. Ugrinov A, Sevov SC (2003) Inorg Chem 42:5789
9. Ugrinov A, Sevov SC (2005) Comptes Rendus Chimie 8:1878
10. Guloy AM, Ramlau R, Tang Z, Schnelle W, Baitinger M, Grin Y (2006) Nature 443:320
11. Kauzlarich SM (1996) Chemistry, Structure and Bonding of Zintl Phases and Ions. VCH Publishers, New York
12. Schafer H, Eisenmann B, Müller W (1973) Angew Chem Int Ed 12:694
13. Parthe E (1973) Acta Cryst B 29:2808
14. Currao A, Curda J, Nesper R (1996) Z Anorg Allg Chem 622:85
15. Nesper R (1990) Prog Solid State Chem 20:1
16. Ponou S, Fässler TF (2007) Z Anorg Allg Chem 633:393
17. Queneau V, Sevov SC (1997) Angew Chem Int Ed 36:1754
18. Hoch C, Wendorff M, Röhr C (2002) Acta Cryst C 58:I45
19. Queneau V, Sevov SC (1998) Inorg Chem 37:1358
20. Bobev S, Sevov SC (2002) Polyhedron 21:641
21. Todorov E, Sevov SC (1998) Inorg Chem 37:3889
22. Hoch C, Wendorff M, Röhr C (2003) Z Anorg Allg Chem 629:2391
23. Hoch C, Wendorff M, Röhr C (2003) J Alloys Compd 361:206
24. Queneau V, Todorov E, Sevov SC (1998) J Am Chem Soc 120:3263
25. Carrillo-Cabrera W, Gil RC, Somer M, Persil O, von Schnering HG (2003) Z Anorg Allg Chem 629:601
26. von Schnering HG, Baitinger M, Bolle U, CarrilloCabrera W, Curda J, Grin Y, Heinemann F, Llanos J, Peters K, Schmeding A, Somer M (1997) Z Anorg Allg Chem 623:1037
27. Pauling L (1960) The nature of the chemical bond and the structure of molecules and crystals. Cornell University Press, Ithaca
28. Pearson RG (1995) Inorg Chim Acta 240:93
29. Evers J, Oehlinger G, Weiss A (1977) Angew Chem Int Ed 16:659
30. Evers J, Oehlinger G, Weiss A (1977) Z Naturforsch B 32:1352
31. Evers J, Oehlinger G, Weiss A (1977) J Solid State Chem 20:173
32. Witte J, von Schnering HG (1964) Z Anorg Allg Chem 327:260
33. Goebel T, Prots Y, Haarmann F (2008) Z Kristallogr NCS 223:187
34. Busmann E (1961) Z Anorg Allg Chem 313:90
35. von Schnering HG, Schwarz M, Chang J-H, Peters K, Peters EM, Nesper R (2005) Z Kristallogr NCS 220:525
36. von Schnering HG, Schwarz M, Nesper R (1986) Angew Chem 98:558
37. Janzon KH, Schafer H, Weiss A (1970) Z Anorg Allg Chem 372:87
38. Eisenmann B, Riekel C, Schafer H, Weiss A (1970) Z Anorg Allg Chem 372:325
39. von Schnering HG, Llanos J, Chang JH, Peters K, Peters EM, Nesper R (2005) Z Kristallogr NCS 220:324
40. Llanos J, Nesper R, von Schnering HG (1983) Angew Chem Int Ed 22:998
41. von Schnering HG, Llanos J, Grin Y, Carrillo-Cabrera W, Peters EM, Peters K (1998) Z Kristallogr NCS 213
42. Betz A, Schafer H, Weiss A, Wulf R (1968) Z Naturforsch B 23:878
43. Müller W, Volk K (1977) Z Naturforsch B 32:709
44. Grin Y, Baitinger M, Kniep R, von Schnering HG (1999) Z Kristallogr NCS 214:453
45. Hewaidy IF, Busmann E, Klemm W (1964) Z Anorg Allg Chem 328:283
46. Baitinger M, Grin Y, von Schnering HG, Kniep R (1999) Z Kristallogr NCS 214:457
47. Marsh RE, Shoemaker DP (1953) Acta Cryst 6:197
48. Röhr C (1995) Z Naturforsch B 50:802
49. Baitinger M, Peters K, Somer M, Carillo-Cabrera W, Grin Y, Kniep R, von Schnering HG (1999) Z Kristallogr NCS 214:455
50. Bobev S, Sevov SC (2000) Angew Chem Int Ed 39:4108
51. Wade K (1976) Adv Inorg Radiochem 18
52. Williams RE (1992) Chem Rev 92:177

53. Mingos DMP (1972) Nature 236:99
54. Bobev S, Sevov SC (2002) J Am Chem Soc 124:3359
55. Bobev S, Sevov SC (2001) Inorg Chem 40:5361
56. Bobev S, Sevov SC (2001) Angew Chem Int Ed 40:1507
57. Eisenmann B, Janzon KH, Schaefer H, Weiss A (1969) Z Naturforsch B 24:457
58. Aydemir U, Ormeci A, Borrmann H, Bohme B, Zurcher F, Uslu B, Goebel T, Schnelle W, Simon P, Carrillo-Cabrera W, Haarmann F, Baitinger M, Nesper R, von Schnering HG, Grin Y (2008) Z Anorg Allg Chem 634:1651
59. Zurcher F, Nesper R (1998) Angew Chem Int Ed 37:3314
60. Nesper R, von Schnering HG, Curda J (1986) Chem Ber 119:3576
61. von Schnering HG, Nesper R, Curda J, Tebbe KF (1980) Angew Chem Int Ed 19:1033
62. Nesper R, Curda J, von Schnering HG (1986) J Solid State Chem 62:199
63. Frank U, Müller W (1975) Z Naturforsch 30:313
64. Todorov I, Sevov SC (2005) Inorg Chem 44:5361
65. Todorov I, Sevov SC (2004) Inorg Chem 43:6490
66. von Schnering HG, Bolle U, Curda J, Peters K, CarrilloCabrera W, Somer M, Schultheiss M, Wedig U (1996) Angew Chem Int Ed 35:984
67. Bolle U, Carrillo-Cabrera W, Peters K, von Schnering HG (1998) Z Kristallogr NCS 213:689
68. Nesper R, Currao A, Wengert S (1998) Chem Eur J 4:2251
69. Eisenmann B, Schafer H, Turban K (1974) Z Naturforsch B 29:464
70. Corbett JD (1985) Chem Rev 85:383
71. Savin A, Becke AD, Flad J, Nesper R, Preuss H, von Schnering HG (1991) Angew Chem Int Ed 30:409
72. Becke AD, Edgecombe KE (1990) J Chem Phys 92:5397
73. Rieger W, Parthe E (1967) Acta Cryst 22:919
74. Nesper R, Wengert S (1999) Chem Monthly 130:197
75. Wengert S, Nesper R (1998) Z Anorg Allg Chem 624:1801
76. Wengert S, Nesper R (2000) Inorg Chem 39:2861
77. Zurcher F, Nesper R (2001) Z Kristallogr NCS 216:507
78. Palenzona A, Manfrinetti P, Fornasini ML (2002) J Alloys Compd 345:144
79. Zurcher F, Nesper R (2001) Z Kristallogr NCS 216:505
80. Müller W (1974) Z Naturforsch B 29:304
81. Palenzona A, Manfrinetti P, Fornasini ML (2000) J Alloys Compd 312:165
82. Nesper R, Currao A, Wengert S (1996) Silicon frameworks and Electronic structures of novel solid silicides. In: Auner N, Weis J (eds) Organosilicon Chemistry. Wiley-VCH, pp 469–491
83. van Leuken H, de Wijs GA, van der Lugt W, de Groot RA (1996) Phys Rev B 53:10599
84. Nesper R, Wengert S, Zurcher F, Currao A (1999) Chem Eur J 5:3382
85. Stalder ED, Wörle M, Nesper R (2010) Inorg Chim Acta 363:4355
86. von Schnering HG, Hönle W (1988) Chem Rev 88:243
87. Wichelhaus W, von Schnering HG (1973) Naturwissenschaften 60:104
88. von Schnering HG, Somer M, Kliche G, Hönle W, Meyer T, Wolf J, Ohse L, Kempa PB (1991) Z Anorg Allg Chem 601:13
89. Emmerling F, Röhr C (2003) Z Anorg Allg Chem 629:467
90. Manriquez V, Hönle W, von Schnering HG (1986) Z Anorg Allg Chem 539:95
91. Santandrea RP, Mensing C, von Schnering HG (1986) Thermochim Acta 98:301
92. Hönle W, Meyer T, Mensing C, von Schnering HG (1985) Z Kristallogr 170:78
93. Meyer T, Hönle W, von Schnering HG (1987) Z Anorg Allg Chem 552:69
94. Dahlmann W, von Schnering HG (1972) Naturwissenschaften 59:420
95. Dahlmann W, Schnerin HG (1973) Naturwissenschaften 60:429
96. Emmerling F, Röhr C (2002) Z Naturforsch B 57:963
97. Hönle W, Buresch J, Peters K, Chang JH, von Schnering HG (2002) Z Kristallogr NCS 217:485

98. Hönle W, Buresch J, Peters K, Chang JH, von Schnering HG (2002) Z Kristallogr NCS 217:487
99. Hönle W, Buresch J, Wolf J, Peters K, Chang JH, von Schnering HG (2002) Z Kristallogr NCS 217:489
100. Schmettow W, von Schnering HG (1977) Angew Chem Int Ed 16:857
101. Dorn FW, Klemm W (1961) Z Anorg Allg Chem 309:189
102. Hirschle C, Röhr C (2000) Z Anorg Allg Chem 626:1992
103. Hackspill L, Bossuet R (1912) C R Acad Sci 154:202
104. Royen P, Zschaage W, Wutschel A (1955) Angew Chem 67:75
105. von Schnering HG (1973) Nachr Chem Techn 21:440
106. Abicht HP, Hönle W, von Schnering HG (1984) Z Anorg Allg Chem 519:7
107. von Schnering HG, Meyer T, Hönle W, Schmettow W, Hinze U, Bauhofer W, Kliche G (1987) Z Anorg Allg Chem 553:261
108. Schmettow W, Lipka A, von Schnering HG (1974) Angew Chem Int Ed 13:345
109. Hönle W, Krogull G, Peters K, von Schnering HG (1999) Z Kristallogr NCS 214:17
110. Emmerling F, Petri D, Röhr C (2004) Z Anorg Allg Chem 630:2490
111. Burns RC, Gillespie RJ, Barnes JA, McGlinchey MJ (1982) Inorg Chem 21:799
112. Kraus F, Hanauer T, Korber N (2005) Angew Chem Int Ed 44:7200
113. Kraus F, Schmedt auf der Günne J, DiSalle BF, Korber N (2006) Chem Commun 218
114. von Schnering HG, Hartweg M, Hartweg U, Hönle W (1989) Angew Chem Int Ed 28:56
115. Derrien G, Tillard M, Manteghetti A, Belin C (2003) Z Anorg Allg Chem 629:1601
116. von Schnering HG, Wittmann M, Sommer D (1984) Z Anorg Allg Chem 510:61
117. Eisenmann B, Rossler U (2003) Z Anorg Allg Chem 629:459
118. Gascoin F, Sevov SC (2001) Inorg Chem 40:5177
119. Deller K, Eisenmann B (1976) Z Naturforsch B 31:1023
120. Deller K, Eisenmann B (1977) Z Naturforsch B 32:1368
121. Somer M, Hartweg M, Peters K, von Schnering HG (1991) Z Kristallogr 195:103
122. Eisenmann B, Jordan H, Schafer H (1985) Z Naturforsch B 40:1603
123. Hadenfeldt C, Bartels F (1994) Z Anorg Allg Chem 620:1247
124. Hönle W, Miller G, Simon A (1988) J Solid State Chem 75:147
125. Fässler TF (2001) Coord Chem Rev 215:347
126. Sevov SC, Goicoechea JM (2006) Organometallics 25:4530
127. Joseph S, Suchenrunk C, Kraus F, Korber N (2009) Eur J Inorg Chem 2009:4641
128. Joseph S, Suchentrunk C, Korber N (2010) Z Naturforsch B65:1059
129. Suchentrunk C, Daniels J, Somer M, Carrillo-Cabrera W, Korber N (2005) Z Naturforsch B60:277
130. Somer M, Carrillo-Cabrera W, Peters EM, Peters K, von Schnering HG (1998) Z Anorg Allg Chem 624:1915
131. Carrillo-Cabrera W, Aydemir U, Somer M, Kircali A, Fässler TF, Hoffmann SD (2007) Z Anorg Allg Chem 633:1575
132. Diehl L, Khodadadeh K, Kummer D, Strähle J (1976) Z Naturforsch B 31:522
133. Diehl L, Khodadadeh K, Kummer D, Strähle J (1976) Chem Ber 109:3404
134. Korber N, Fleischmann A (2001) Dalton Trans 2001:383
135. Wiesler K, Brandl K, Fleischmann A, Korber N (2009) Z Anorg Allg Chem 635:508
136. Goicoechea JM, Sevov SC (2004) J Am Chem Soc 126:6860
137. Suchentrunk C, Korber N (2006) New J Chem 30:1737
138. Campbell J, Schrobilgen GJ (1997) Inorg Chem 36:4078
139. Somer M, Carrillo-Cabrera W, Peters EM, Peters K, Kaupp M, von Schnering HG (1999) Z Anorg Allg Chem 625:37
140. Edwards PA, Corbett JD (1977) Inorg Chem 16:903
141. Corbett JD, Edwards PA (1975) Chem Commun 1975:984
142. Rudolph RW, Wilson WL, Parker F, Taylor RC, Young DC (1978) J Am Chem Soc 100:4629

143. Rosdahl J, Fässler TF, Kloo L (2005) Eur J Inorg Chem 2005:2888
144. Korber N, Daniels J (1996) Helv Chim Acta 79:2083
145. Kraus F, Korber N (2005) Chem Eur J 11:5945
146. Korber N, Daniels J (1999) Z Anorg Allg Chem 625:189
147. Korber N, Daniels J (1996) Dalton Trans 1653
148. Korber N, Daniels J (1996) Acta Cryst C 52:2454
149. Korber N, von Schnering HG (1996) Chem Ber 129:155
150. Hönle W, von Schnering HG, Schmidpeter A, Burget G (1984) Angew Chem Int Ed 23:817
151. Knettel D, Reil M, Korber N (2001) Z Naturforsch B56:965
152. Korber N, Daniels J (1996) Z Anorg Allg Chem 622:1833
153. Korber N, Daniels J (1996) Polyhedron 15:2681
154. Korber N, Daniels J, von Schnering HG (1996) Angew Chem Int Ed 35:1107
155. Dai FR, Xu L (2007) Chin J Struct Chem 26:45
156. Hanauer T, Grothe M, Reil M, Korber N (2005) Helv Chim Acta 88:950
157. Somer M, Hönle W, von Schnering HG (1989) Z Naturforsch B 44:296
158. Driess M, Merz K, Pritzkow H, Janoschek R (1996) Angew Chem Int Ed 35:2507
159. Bashall A, Beswick MA, Choi N, Hopkins AD, Kidd SJ, Lawson YG, Mosquera MEG, McPartlin M, Raithby PR, Wheatley A, Wood JA, Wright DS (2000) Dalton Trans 479
160. Hubler K, Becker G (1998) Z Anorg Allg Chem 624:483
161. Korber N, von Schnering HG (1997) Z Kristallogr NCS 212:85
162. Castleman AW, Khanna SN, Sen A, Reber AC, Qian M, Davis KM, Peppernick SJ, Ugrinov A, Merritt MD (2007) Nano Lett 7:2734
163. Hanauer T, Korber N (2006) Z Anorg Allg Chem 632:1135
164. Belin CHE (1980) J Am Chem Soc 102:6036
165. Adolphson DG, Corbett JD, Merryman DJ (1976) J Am Chem Soc 98:7234
166. Beswick MA, Choi N, Harmer CN, Hopkins AD, McPartlin M, Wright DS (1998) Science 281:1500
167. Critchlow SC, Corbett JD (1984) Inorg Chem 23:770
168. Breunig HJ, Ghesner ME, Lork E (2005) Z Anorg Allg Chem 631:851
169. Kraus F, Hanauer T, Korber N (2006) Inorg Chem 45:1117
170. Kraus F, Aschenbrenner JC, Korber N (2003) Angew Chem Int Ed 42:4030
171. Hanauer T, Aschenbrenner JC, Korber N (2006) Inorg Chem 45:6723
172. Korber N (1997) Phosphorus Sulfur Silicon 125:339
173. Hanauer T, Kraus F, Reil M, Korber N (2006) Chem Monthly 137:147
174. Haushalter RC, Eichhorn BW, Rheingold AL, Geib SJ (1988) Chem Commun 1027
175. Korber N, Richter F (1997) Angew Chem Int Ed 36:1512
176. Reil M, Korber N (2007) Z Anorg Allg Chem 633:1599
177. Bolle U, Tremel W (1992) Chem Commun 91
178. Baudler M, Glinka K (1993) Chem Rev 93:1623
179. Baudler M, Glinka K (1994) Chem Rev 94:1273
180. Kraus F, Aschenbrenner JC, Klamroth T, Korber N (2009) Inorg Chem 48:1911
181. Korber N, Aschenbrenner J (2001) Dalton Trans 1165
182. Dai F-R, Xu L (2006) Inorg Chim Acta 359:4265
183. Aschenbrenner AC, Korber N (2004) Z Anorg Allg Chem 630:31
184. Korber N, von Schnering HG (1995) Chem Commun 1713
185. Ye YZ, Xu L (2008) Chin J Struct Chem 27:75
186. Korber N, Daniels J (1997) Inorg Chem 36:4906
187. Ogg RAJ (1954) J Chem Phys 22:560
188. Jolly WL (1952) Chem Rev 50:351
189. Franklin EC (1905) J Am Chem Soc 27:831

Struct Bond (2011) 140: 59–89
DOI: 10.1007/430_2010_23

Published online: 27 October 2010

Dynamic Properties of the Group 14 Zintl Ions and Their Derivatives

Bryan Eichhorn and Sanem Kocak

Abstract A complete review of the NMR, EPR, general spectroscopic properties, and the dynamic behavior of the group 14 Zintl ions and their derivatives are provided. The review focuses primarily on the ^{207}Pb and ^{119}Sn NMR studies involving derivatives of the E_9^{4-} ions (E = Sn, Pb) and their coupled products. These clusters and their transition metal derivatives show a wide range of dynamic behavior, including rapid global exchange of up to 18 metal atoms in the $Pd_2@Sn_{18}^{4-}$ ion. The compounds span large spectral regions of the chemical shift windows ($\Delta\delta$ ^{119}Sn = 4,388 ppm, $\Delta\delta$ ^{207}Pb = 5,965 ppm) and show a variety of couplings to various heteroatoms (e.g., ^{1}H, ^{31}P, ^{65}Cu, ^{195}Pt). In addition, ^{117}Sn satellites on ^{119}Sn NMR resonances are used to assess fluxionality and nuclearity in the tin-containing derivates. Correlations of chemical shifts, coupling constants, and dynamic behavior are presented in connection with three-dimensional structures and mechanisms of dynamic exchange. The fluxionality of the group 14 metal clusters and the substrate mobilities of the attendant substituents (i.e., H, R, SnR_3) have particular relevance to small bimetallic nanoparticle catalysts. The relationship between the bimetallic Zintl clusters and atomic mobilities in heterogeneous catalytic processes is described.

Keywords Clusters · Dynamic NMR · ^{119}Sn NMR · Zintl ions

Contents

B. Eichhorn (✉) and S. Kocak
Department of Chemistry and Biochemistry, University of Maryland, College Park, MD 20742, USA
e-mail: eichhorn@umd.edu

1 Introduction

In this chapter, we present a complete review of the NMR spectroscopic properties and the dynamic behavior of the group 14 Zintl ions and their derivatives. Zintl ions, such as Sn_9^{4-}, are anionic bare metal clusters of the main group elements that are soluble in polar, aprotic solvents. These clusters and their transition metal derivatives show a wide range of dynamic behavior; including rapid global exchange of up to 18 metal atoms. The mobility of the constituent metal atoms and the attendant groups provides insight into the stability and activity of small metallic nanoparticles and bimetallic catalysts.

The group 14 Zintl ions are characterized by electron-deficient multicenter bonding that is reminiscent of the borane and carborane cages. Detailed discussions of their structures and bonding can be found in the other chapters of this volume. Importantly, the element–element bond energies of the heavier elements in this group (e.g., Pb, Sn, Ge) are approximately 50% less than those of boron or carbon. Weak bonds and delocalized, degenerate states give rise to low energy barriers for atomic rearrangement (i.e., high fluxionality) among the group 14 Zintl complexes. The presence of spin-active I = ½ nuclei in this group (^{207}Pb, ^{119}Sn, ^{117}Sn, and potentially ^{29}Si) facilitates multinuclear NMR studies of this class of fluxional clusters. The vast majority of this review focuses on ^{119}Sn and ^{207}Pb NMR studies. Overviews of their NMR properties are given below.

Tin has nine naturally occurring isotopes but only two have spin-active nuclei (^{119}Sn, 8.7% abund., I = ½; ^{117}Sn, 7.7% abund., I = ½; remaining isotopes ^{X}Sn, 83.6% abund., I = 0). Because ^{119}Sn is slightly more abundant and has a slightly higher sensitivity, it is typically the isotope of choice for NMR investigations. Using an 11.744 Tesla magnet (i.e., a "500 MHz" NMR), the resonance frequency of ^{119}Sn is 186.4 MHz. The ^{119}Sn chemical shift window is quite large and the Zintl clusters described herein appear in the range of −1,895 to 2,493 ppm. The ^{119}Sn chemical shifts (δ ^{119}Sn) are referenced to $SnMe_4$ with δ ^{119}Sn = 0.0 ppm.

Lead has four naturally occurring isotopes with only one spin-active nucleus (^{207}Pb, 21.1% abund., I = ½; remaining isotopes ^{X}Pb, 78.9% abund., I = 0). With an 11.744 Tesla magnet (i.e., a "500 MHz" NMR), the resonance frequency of ^{207}Pb is 104.6 MHz. The ^{207}Pb NMR chemical shift window is enormous and the compounds described herein span the range from −4,185 to +1,780 ppm relative to $PbMe_4$. Note that some of the older literature references ^{207}Pb NMR chemical shifts (δ ^{207}Pb) relative to 1.0 M $Pb(NO_3)_2$ in D_2O. The currently accepted primary standard is $PbMe_4$ (δ ^{207}Pb = 0.0 ppm) which gives 1.0 M $Pb(NO_3)_2$ in D_2O a chemical shift of −2,961 ppm. In this review, the older data have been adjusted to reflect the modern $PbMe_4$ standard.

This review will focus on the NMR properties of Zintl ion complexes, namely the solution properties of the E_9^{4-} ions where E = Si, Ge, Sn, Pb, and the products derived from those clusters. Closely related clusters prepared by other means, such as the recent, elegant organo polystannane work of Schnepf, Power, Huttner, and Fischer, are briefly mentioned but are not the focus of this review. Related overviews of dynamic organometallic complexes [1, 2] and the structure and bonding of Zintl ions [3–5] can be found in previous reviews and in other chapters of this book.

1.1 Relevance

The mobility of metal atoms in bare metal clusters and small metallic nanoparticles (NPs) is of fundamental importance to cluster science and nanochemistry. Atomic mobility also has significant implications in the reactivity of catalysts in heterogeneous transformation [6]. Surface restructuring in bimetallic NP and cluster catalysts is particularly relevant because changes in the local environment of a metal atom can alter its chemical activity [7, 8].

Surface-to-subsurface migrations in bimetallic NPs can completely change the surface composition and drastically alter catalytic performance. Despite the importance of atomic mobility in bimetallic systems, little is known about the dynamic processes of NPs and clusters due, in large part, to the lack of suitable experimental techniques.

Recent studies suggest that small NPs are far more fluid than initially thought. Lower melting points [9] and the dominance of surface energies [10] in small NPs suggest that these systems may be quite dynamic, which can be used to make new nanomaterials. By exploiting the differences in adsorbate bond enthalpies to different metals in small bimetallic NPs [8, 11], randomly mixed alloys can be converted to core–shell NPs near room temperature [12–14]. Recent in situ XPS studies by Tao et al. have shown that surface-to-subsurface migrations of CO and Pt atoms occur quickly at room temperature and are highly reversible [10]. In addition, the transport of reactants (e.g., H, CO, and hydrocarbyl) across a catalyst surface is central to their catalytic activity but is not well understood in NPs and bare metal clusters.

Inorganic transition metal clusters have long been used as models for metallic catalysts, but the comparisons are hampered by the presence of strongly bound ligands (e.g., phosphines, CO, and amides) on the cluster compounds that significantly alter the structures and reactivities of the metal cores. In contrast, Zintl ions have no attendant ligands and can be considered to be very small metal nanoparticles with a few extra electrons. As such, they are excellent precursors to well-defined nanomaterials [15] and serve as excellent models for studying particle properties. In particular, the mobility of the metal atoms in the bimetallic Zintl clusters and the migration of hydrogen and other attendant substituents provide unique insight into the structural chemistry of bimetallic clusters. Because tin and lead have naturally occurring, NMR spin-active I = ½ nuclei (^{119}Sn, ^{117}Sn, ^{207}Pb), the dynamic properties are ideally suited for study by NMR spectroscopy. The coupled Ge_9 clusters have been the most well-studied Zintl clusters to date and provide the best structural link to nanomaterials [16–21]. Unfortunately, the only spin-active isotope of germanium (^{73}Ge, I = 9/2) is quadrupolar, and therefore the fluxionality of these compounds remains unknown.

1.2 History

The possibility of dynamic exchange in the group 14 Zintl ions was first proposed in 1977 by Corbett and Edwards in their structural and theoretical study of Sn_9^{4-} [22]. The observed C_{4v} structure of Sn_9^{4-} differed from the expected D_{3h} structure (Fig. 1) of the isoelectronic Bi_9^{5+} cation, which led to the postulate that the two geometries were very close in energy and most likely in equilibrium in solution. SCF-CNDO-MO calculations of the limiting C_{4v} and D_{3h} Sn_9^{4-} structures suggested that the energy difference between the two was negligible.

In 1978, Rudolph and coworkers provided the first experimental evidence for dynamic exchange in their landmark ^{119}Sn NMR studies of a melt having the nominal composition $NaSn_{2.25}$ (i.e., Na_4Sn_9) dissolved in ethylenediamine (en) [23]. These solutions were known to contain Sn_9^{4-} ions, but direct observation of these species had not been achieved before these experiments. Because the ^{119}Sn chemical shift window is quite large and the fact that multinuclear FT NMR spectroscopy was in its infancy in 1977, Rudolph used $NaSn_{2.25}$ melts that were isotopically enriched in ^{119}Sn (85%) to locate the signal for Sn_9^{4-}. After finding the single peak at −1,230 ppm (1,230 ppm upfield of tetramethyltin), they recorded the ^{119}Sn NMR spectrum of the natural abundance sample (^{119}Sn, 8.7%, I = ½; ^{117}Sn, 7.7%, I = ½) at both 31.896 and 29.641 MHz. Instead of finding three signals in

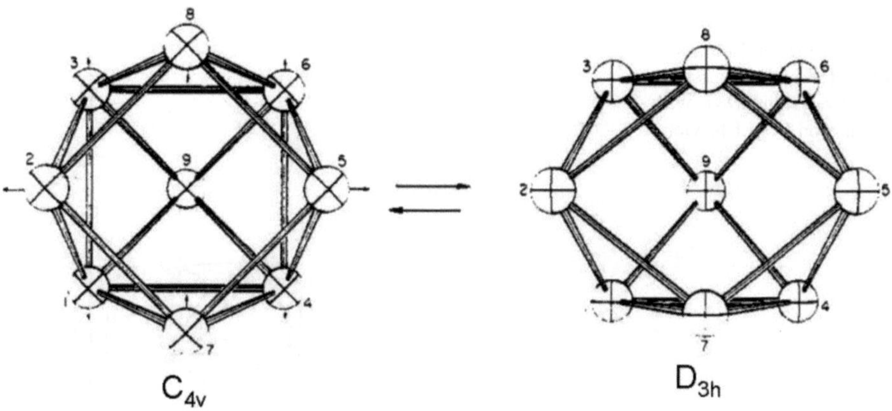

Fig. 1 Scheme of the interconversion between two lowest energy symmetries for the E_9^{4-} Zintl ions. Reprinted with permission [22]. Copyright 1977 American Chemical Society

4:4:1 ratio as expected from the C_{4v} solid-state structure, the Sn_9^{4-} ion showed a single resonance in en and NH_3 solvents from 30°C to −40°C due to rapid intramolecular exchange of all nine Sn atoms.

The intramolecular nature of exchange and the nuclearity of the cluster are conveniently evidenced by the presence and intensities of the ^{117}Sn satellites on the ^{119}Sn resonance. The fortuitous presence of almost equal abundance of I = ½ ^{119}Sn and ^{117}Sn isotopes provides a direct measure of the number of Sn atoms in a given cluster. A statistical evaluation of the isotope distributions in the clusters gives the relative populations the different isotopimers (i.e., clusters having different distributions of isotopes). The multiplicity of the resonance associated with each isotopimer (i.e., singlet, doublet, triplet, etc.) and its relative abundance dictates the satellite pattern and intensity. The ^{119}Sn–^{117}Sn coupling is always observed, even between equivalent atoms or inequivalent atoms involved in fast exchange, whereas first-order ^{119}Sn–^{119}Sn coupling is only observed between structurally unique atoms that are static on the NMR time scale. Rudolph exploited this unique property of tin to show that Sn_9^{4-} was a dynamic nine-atom cluster and was the prominent (only) species in solution. The original spectrum and tabulated cluster intensities are given in Fig. 2. Rudolph also reported the ^{207}Pb NMR spectrum for Pb_9^{4-}, which consisted of a single peak at −4,150 ppm [1,190 ppm upfield from 1 M Pb $(NO_3)_2$] [23].

Despite working in the field for only 4 years (1978–1981), Ralph Rudolph made seminal contributions to the field of dynamic exchange in Zintl ions through multinuclear NMR studies. In addition to showing that the E_9^{4-} ions (E = Ge, Sn, Pb) were all highly dynamic in solution, he also showed that main group and transition metal derivatives of the E_9^{4-} ions could be prepared. The remainder of the review is organized along these lines: the naked group 14 polyanions, the main group adducts of the E_9^{4-} ions, and the transition metal derivatives of the group 14 polyanions [25, 26].

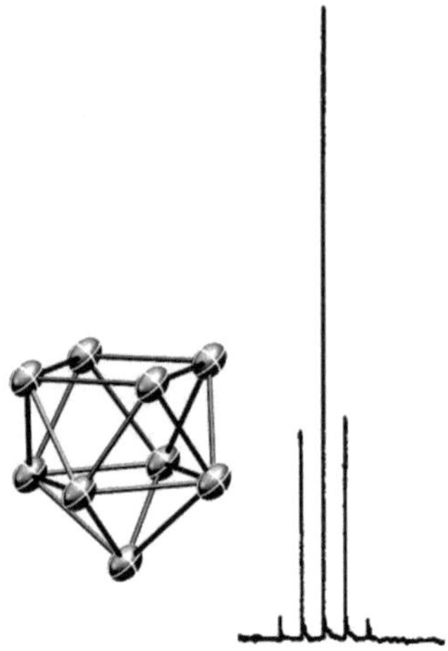

Fig. 2 Structure and ^{119}Sn NMR of the Sn_9^{4-} anion. Reprinted with permission [24]. Copyright 1981, with permission from Elsevier

2 Dynamic Properties of the $E_9^{4-/3-}$, $E'_{9-x}E_x^{4-}$, and E_4^{2-} Ions (E = Pb, Sn, Ge, Si)

A tabular listing of the ^{119}Sn and ^{207}Pb NMR data for the known Zintl complexes Sn_4^{2-}, Sn_9^{4-}, Pb_9^{4-}, and the mixed element analogs (i.e., $Sn_{9-x}Ge_x^{4-}$) are given in Tables 1 and 2. Selected data for various counter cations (i.e., Na, K, etc.) are also given. Full listings can be found in [23, 28]. The spectra of the naked E_{10}^{2-} and E_{12}^{2-} clusters have not been reported as they are only known in the gas phase or, in the case of Pb_{10}^{2-}, only recently synthesized [29]. The transition metal derivatives of these clusters will be discussed in a later section.

2.1 *The E_9^{4-} Ions*

The studies of Rudolph showed that both Sn_9^{4-} and Pb_9^{4-} are dynamic on the NMR time scale, even in liquid NH_3 solutions at −40°C, and give rise to a single resonance in their respective NMR spectra. The ^{119}Sn signal of the Sn_9^{4-} cluster has line intensities which are consistent with the expected intensity distribution of 0.05:0.23:0.65:1:0.65:0.23:0.05 for the 9-equivalent Sn nuclei. The exchange mechanism is widely accepted to involve the interconversion of the D_{3h} and C_{4v} structures by way of the "diamond–square–diamond" mechanism common to the boranes (Fig. 1) [30]. Corbett showed that the energy difference between the two structures was negligible, and more recent studies have shown the calculated energy barrier for interconversion to be <0.01 eV in Sn_9^{4-} [31]. Low temperature ^{119}Sn Mössbauer studies of $(Na\text{-}2,2,2\text{-crypt})_4Sn_9$ crystals at 77 K showed three

Table 1 ^{119}Sn NMR data for the group 14 Zintl ions[a, b]

Cluster ion	Cation	δ^{119}Sn	$J_{119Sn-117Sn}$	Cluster ion	Cation	δ^{119}Sn	$J_{119Sn-117Sn}$
Sn_4^{2-}	Na	−1895	1224	Sn_8Pb^{4-}	Li	−1281	288
Sn_9^{4-}	Li	−1241	273		Na	−1272	262
	Na in NH_3	−1308	257		K	−1251	275
	Na.crypt	−1230	256		Rb	−1231	282
	K	−1210	263		Cs	−1152	303
	K in dmf	−1149	303	$Sn_7Pb_2^{4-}$	Na	−1319	281
	Rb	−1190	268		K	−1298	293
	Cs	−1115	293		Rb	−1276	301
Sn_8Ge^{4-}	Na	−1223	258		Cs	−1194	325
	K	−1204	268	$Sn_6Pb_3^{4-}$	Na	−1366	317
	Rb	−1183	275		K	−1344	323
$Sn_7Ge_2^{4-}$	Na	−1201	264		Rb	−1321	340
	K	−1192	278		Cs	−1235	369
	Rb	−1161	284	$Sn_5Pb_4^{4-}$	Na	−1420	366
$Sn_6Ge_3^{4-}$	Na	−1185	298		K	−1397	378
	K	−1167	309		Rb	−1373	393
	Rb	−1145	317		Cs	−1281	418
$Sn_5Ge_4^{4-}$	Na	−1180	394	$Sn_4Pb_5^{4-}$	Na	−1480	427
	K	−1161	412		K	−1456	455
	Rb	−1139	416		Rb	−1430	469
$Sn_4Ge_5^{4-}$	Na	−1187	494		Cs	−1333	534
	K	−1168	518	$Sn_3Pb_6^{4-}$	Na	−1538	520
	Rb	−1145	518		K	−1512	–
$Sn_3Ge_6^{4-}$	Na	−1188	682		Rb	−1484	584
	K	−1167	713		Cs	−1381	610
	Rb	−1144	726	$Sn_2Pb_7^{4-}$	Na	−1579	–
$Sn_2Ge_7^{4-}$	Na	−1202	–		K	−1553	–
	K	−1178	747		Rb	−1523	717
$SnGe_8^{4-}$	Na	−1237	–		Cs	−1414	753
	K	−1214	–	$SnPb_8^{4-}$	Na	−1610	–
					K	−1587	–
					Rb	−1554	–
					Cs	−1438	–

[a] All spectra were recorded from ethylenediamine solutions unless otherwise indicated
[b] Data were taken from [23, 27]. Chemical shifts are relative to Me_4Sn

environments in roughly a ~4:4:1 ratio, which is consistent with the C_{4v} structure observed in the X-ray analysis [32]. At higher temperatures, the ^{119}Sn Mössbauer studies and related EXAFS studies [33] of Sn_9^{4-} suggest that the cluster shows dynamic behavior in the solid state.

The solution dynamics of the Sn_9^{4-} ion are evidenced not only by the observation of a single NMR chemical shift, but also by the reduced ^{119}Sn–^{117}Sn coupling constants, J(^{119}Sn–^{117}Sn). A typical Sn–Sn single bond in a Zintl complex shows J(^{119}Sn–^{117}Sn) values between 1975 and 1043 Hz [23, 34, 35]. However, the Sn_9^{4-} ion shows J(^{119}Sn–^{117}Sn) values between 250 and 300 Hz, depending on the solvent and the counter cation (see below). These small coupling constants are due to averaging of one-bond, two-bond, and three-bond couplings present in the static structures. Because the dynamic exchange within Sn_9^{4-} remains fast on the NMR

Table 2 ^{207}Pb NMR data for the group 14 Zintl ions[a]

Cluster ion	Cation	δ^{207}Pb	$J_{119Sn-207Pb}$	Cluster ion	Cation	δ^{207}Pb	$J_{119Sn-207Pb}$
Pb_9^{4-}	Li	−4185	–	$Sn_4Pb_5^{4-}$	Na	−3343	568
	Na	−4150	–		K	−3327	550
	K	−4098	–		Rb	−3264	542
	Rb	−4022	–		Cs	−3068	505
	Cs	−3780	–	$Sn_3Pb_6^{4-}$	Na	−3534	562
Sn_8Pb^{4-}	Na	−2667	461		K	−3517	542
	K	−2650	555		Rb	−3452	526
	Rb	−2594	550		Cs	−3245	494
$Sn_7Pb_2^{4-}$	Na	−2825	562	$Sn_2Pb_7^{4-}$	Na	−3731	537
	K	−2808	557		K	−3717	521
	Rb	−2751	546		Rb	−3648	504
	Cs	−2585	520		Cs	−3430	462
$Sn_6Pb_3^{4-}$	Na	−3001	568	$SnPb_8^{4-}$	Na	−3932	488
	K	−2985	557		K	−3911	488
	Rb	−2926	546		Rb	−3848	473
	Cs	−2740	520		Cs	−3609	429
$Sn_5Pb_4^{4-}$	Na	−3168	572				
	K	−3152	557				
	Rb	−3091	546				
	Cs	−2899	517				

[a]Values were taken from [23] and chemical shifts recalibrated against Me_4Pb (δ^{207}Pb = 0.0 ppm) where $Pb(NO_3)_2$ δ^{207}Pb = −2961 ppm

time scale at low temperatures, the J(^{119}Sn–^{117}Sn) values of the limiting structure remain unknown. However, some of the metallated derivatives of the Sn_9^{4-} ion provide insight into the static NMR values and are discussed in a later section of this chapter.

The chemical shifts of Sn_9^{4-} and Pb_9^{4-} ions are both solvent- and cation-dependent. En solutions made from different $ASn_{2.25}$ melts (A = Li, Na, K, Rb, Cs) give rise to different ^{119}Sn NMR chemical shift (δ ^{119}Sn) values (Table 1) that systematically move downfield from A = Li (δ ^{119}Sn = −1,241 ppm) to A = Cs (δ ^{119}Sn = −1,115 ppm). En solutions of the Pb_9^{4-} ions show similar trends of the ^{207}Pb NMR chemical shifts (δ ^{207}Pb) as the alkali cation is varied (Table 2). These trends of downfield shifts with increasing size and “softness” of alkali counter ion are attributed to increased ion pairing with the larger alkali ions [23, 24, 27, 36]. The J(^{119}Sn–^{117}Sn) values are less affected by the nature of the counter cation. Strong ion pairing has also been documented in the polyphosphide Zintl ions, such as Li_3P_7 [37].

The use of 2,2,2-cryptand and various crown ethers greatly reduces ion pairing in solution but affects only minor changes in the chemical shifts of the E_9^{4-} ions. Solvents that reduce K^+–Sn_9^{4-} ion paring through coordination of the alkali metal ion (e.g., en) also shift the δ ^{119}Sn values upfield relative to those solvents that are weakly coordinating (e.g., dmf; Table 1).

While NMR data directly reveal the dynamic nature of the Sn_9^{4-} and Pb_9^{4-} ions, ^{119}Sn NMR data for the $Sn_{9-x}Ge_x^{4-}$ clusters (x = 1→7) show [23] that these clusters are also dynamic (see below), suggesting that Ge_9^{4-} is also highly fluxional. However, the lack of an I = ½ isotope for Ge has precluded direct measurements of its dynamic properties. The ^{29}Si NMR spectrum of the recently synthesized Si_9^{4-} ion has not been reported to date. Although ^{29}Si NMR spectroscopy is readily available (^{29}Si, 59.6%, I = ½), ^{29}Si nuclei often have very long T_1 relaxation times, which may hinder NMR studies of this cluster.

The $Sn_{9-x}Ge_x^{4-}$ clusters (x = 1→8) and $Sn_{9-x}Pb_x^{4-}$ clusters (x = 1→8) have also been characterized by ^{119}Sn and ^{207}Pb NMR spectroscopy [23, 24, 27, 36, 38]. The $E_{9-x}E'_x{}^{4-}$ ions are highly fluxional but also show very interesting systematic trends (Tables 1 and 2). First, the ^{119}Sn chemical shifts of the $Sn_{9-x}Pb_x^{4-}$ clusters move systematically upfield with each increasing Pb atom (~45–60 ppm/Pb atom; Fig. 3). The ^{207}Pb NMR chemical shifts (δ ^{207}Pb) move ~200 ppm/Pb atom upfield in the same series. In contrast, the $Sn_{9-x}Ge_x^{4-}$ clusters do not show a significant correlation between δ ^{119}Sn and x. Second, the J(^{119}Sn–^{117}Sn) coupling constants increase with increasing x in both the $Sn_{9-x}Pb_x^{4-}$ and $Sn_{9-x}Ge_x^{4-}$ series, whereas the J(^{119}Sn–^{207}Pb) coupling constants are relatively constant.

These data indicate that Sn atoms in the $Sn_{9-x}E'_x{}^{4-}$ clusters have a tendency to self-associate instead of being randomly distributed during dynamic exchange. In particular, the increase in J(^{119}Sn–^{117}Sn) values with increasing x suggests that each tin atom experiences a greater frequency of direct, one-bond Sn–Sn interactions (larger J) relative to indirect two- and three-bond interactions (smaller J) [23]. This increase in direct Sn–Sn interactions with decreasing numbers of tin atoms is strongly suggestive of self-association. This self-association would not affect the Sn–Pb interactions and is consistent with the lack of change in the J(^{119}Sn–^{207}Pb)

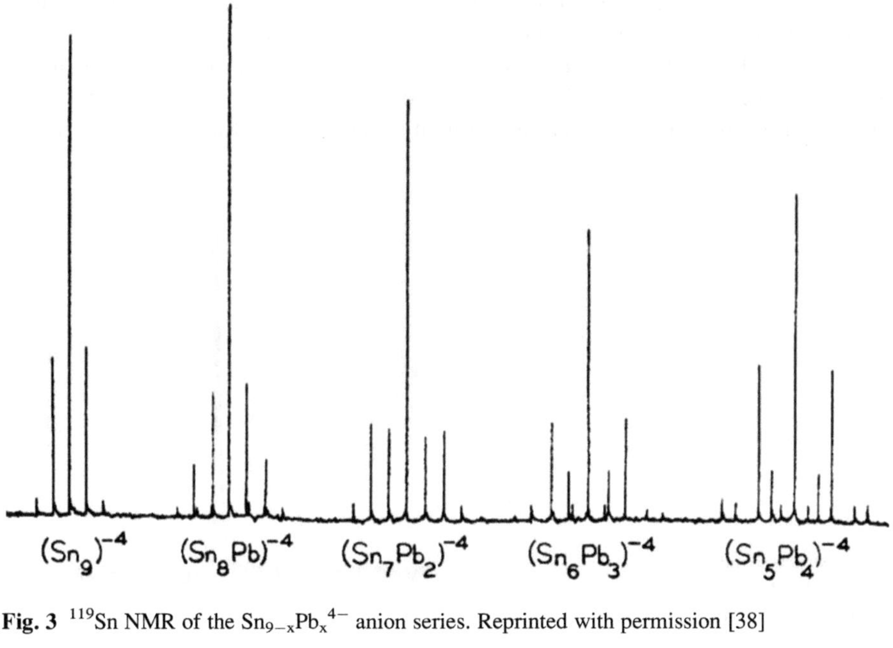

Fig. 3 ^{119}Sn NMR of the $Sn_{9-x}Pb_x^{4-}$ anion series. Reprinted with permission [38]

values in the same series. The origins of the chemical shift changes are less obvious but may be associated with the "heavy atom effect" or differing charge distributions [23]. Finally, as with the pure Sn_9^{4-} and Pb_9^{4-} clusters, the ^{119}Sn and ^{207}Pb chemical shifts of the $Sn_{9-x}E'_x{}^{4-}$ clusters (E = Ge, Pb) move downfield as the counterion is varied from Li^+ to Cs^+; presumably due to greater ion pairing with the heavier alkali ions as described above.

2.2 The E_4^{2-} and E_5^{2-} Ions Where E = Sn, Pb

The E_4^{2-} ions, where E = Sn, Pb, are tetrahedral and therefore NMR spectroscopy cannot be used to probe dynamic behavior (i.e., all atoms are equivalent and symmetrically bonded to the other three atoms) [27, 39]. The ^{119}Sn chemical shift of Sn_4^{2-} (−1,895 ppm) is the most upfield shift for a Zintl complex. The 1,224 Hz ^{119}Sn–^{117}Sn coupling constant is significantly larger that the ~250–310 Hz couplings observed for Sn_9^{4-} but is on the low end of J(^{119}Sn–^{117}Sn) values for single bonded organostannanes (1,500–5,000 Hz) [1]. The chemical shift of the Pb_4^{2-} ion has not been reported.

The NMR spectra of Pb_5^{2-} and Sn_5^{2-} have not been reported even though the clusters were first isolated in 1975 [40]. Preliminary data for Pb_5^{2-} ion show that it is static on the NMR time scale (Fig. 4) [41]. The D_{3h} structure of the trigonal bipyramidal cluster is expected to give rise to two NMR resonances. While only one resonance has been located (δ ^{207}Pb = −3,591 ppm), it clearly shows coupling to a second inequivalent set of Pb atoms with $^1J(^{207}Pb$–^{207}Pb = 2,344 Hz). The calculated intensity distribution for the three equatorial Pb nuclei is expected to be 0.004:0.28:1:0.28:0.004, whereas the signal for the two axial Pb atoms is expected to be 0.054:0.387:1:0.387:0.054. Although it is difficult to compare the observed line intensities for the assignment, experimental data are closer to the

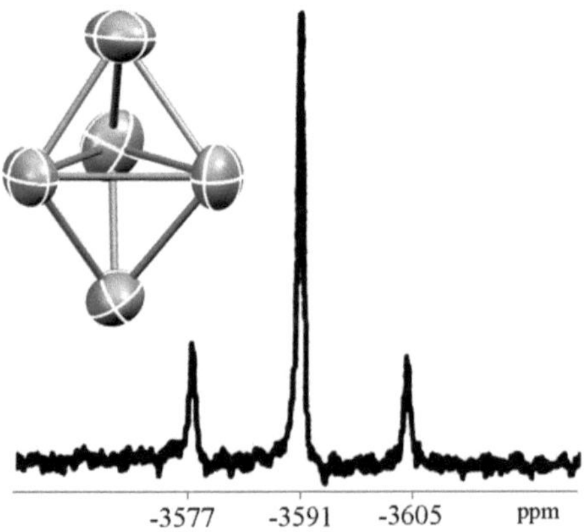

Fig. 4 Structure and ^{207}Pb NMR of the Pb_5^{2-} anion. Reprinted with permission [41]

distribution pattern for the three Pb equatorial nuclei. This coupling indicates that the Pb_5^{2-} cluster is static or is in slow exchange on the NMR time scale.

2.3 The Paramagnetic E_9^{3-} Where E = Ge, Sn, Pb

The paramagnetic E_9^{3-} clusters are known for all group 14 elements except carbon. The solution and solid-state EPR data for these ions are rather broad and do not give well-resolved hyperfine couplings [42]. As such, the data do not provide significant information regarding their fluxionality. However, one can assume that they are similar to the E_9^{4-} ions in their dynamic properties even though the calculated energy difference between the C_{4v} and D_{3h} structures for the E_9^{3-} clusters is slightly higher than in the E_9^{4-} analogs [31].

3 Transition Metal Derivatives of the Group 14 Zintl Ions

The section includes the NMR and EPR data for the transition metal derivatives of the Sn_9^{4-} and Pb_9^{4-} clusters. During the synthesis of these complexes, the E_9^{4-} precursors can fragment, couple, or becomes oxidized to give bimetallic clusters of differing nuclearities and charges. To systematize the discussion of this section, the clusters are discussed in terms of increasing nuclearity of the main group fragments. Summaries of the data are found in Tables 3 and 4.

3.1 The $Sn_6[M(CO)_5]_6^{2-}$ Ions (M = Cr, Mo, W) and $Sn_6Nb_2(toluene)_2^{2-}$

The $Sn_6[M(CO)_5]_6^{2-}$ clusters (M = Cr, Mo, W) [51] are not prepared from Zintl anions, but the Sn_6^{2-} octahedral cores can be viewed as a *closo* 2n+2 member of the deltahedral Zintl anion series and are similar to the *closo*-trigonal bipyramidal Sn_5^{2-} ion. The ^{119}Sn chemical shifts of the $Sn_6[M(CO)_5]_6^{2-}$ complexes are in the range 198–561 ppm and are somewhat downfield of the anionic Zintl clusters. Large one-bond Sn–Sn couplings of ~1,800 Hz are suggestive of nonfluxional compounds.

Unlike the $Sn_6[M(CO)_5]_6^{2-}$ clusters, the $Sn_6Nb_2(toluene)_2^{2-}$ ion contains an open cyclohexane-like Sn_6 ring that can be viewed as an Sn_6^{12-} fragment if all of the Nb valence electrons are partitioned to Sn according to the Zintl formalism [43]. The six Sn atoms are equivalent by symmetry giving rise to a single resonance at $\delta\ ^{119}Sn = -149$ ppm. The structure is not fluxional as evidenced by the magnitude and intensities of the one-bond and two-bond $J(^{119}Sn–^{117}Sn)$ values of 1,975 Hz and 630 Hz, respectively.

Table 3 NMR data for the transition metal derivatives of Tin Zintl ions[a]

A		$\delta^{119}Sn$			$J^{119/117}Sn–^{119/117}Sn$		
Cluster Ion	Metal	Sn6–9	Sn2–5	Sn1	1 and 6–9 1 and 2–5	2–5 and 6–9	Within waists
$Sn_9M(CO)_3^{4-}$	Cr	−180[b]	−447	2327	358–322	975–783	745–129
		−214[c]	−522	2493	1048–		
	Mo	−361[b]	−606	1988	1042		
		−402[c]	−682	2125			
	W	−443[b]	−662	2279			
		−497[c]	−736	2448			
B	X	δ X	$\delta^{119}Sn$	*J*			Ref
				$^{119}Sn–^{117}Sn$	$^{119}Sn–X$	X–X′	
$Sn_6(Nbtol)_2^{2-}$	–	–	−149	1975, 630	–	–	[43]
$CuSn_9^{3-}$	Cu	−330	−1440	85	280	–	[44]
$Sn_9Ir(cod)^{3-}$	–	–	−1149	139	–	–	[45]
$Ni_2Sn_{17}^{4-}$	–	–	−1176	–	–	–	[46]
$Pd_2Sn_{18}^{4-}$	–	–	−734	<70	–	–	[47]
$Pt_2Sn_{17}^{4-}$	Pt	−5270	−742	170	780	–	[48]
$PtSn_9Pt(PPh_3)^{2-}$	Pt	−6010	−862	<150	–	2,472	[49]
	Pt′	−5270			1690		
	P	35.4			–	305, 4800	
$NiSn_9H^{3-}$	H	−28.3	−837	59	51	–	[50]
$PdSn_9H^{3-}$	H	−24.3	−326	~40	43	–	
$PtSn_9H^{3-}$	Pt	−5700	−367	~46	1540	32	[48]
	H	−22.3			46		
$Na_xL_2Pd_ySn_9$	–	–	−755	39	–	–	[26]
$K_xL_2Pt_ySn_9$	Pt	–	−753	95	1554	–	[25]
$Sn_6\{Cr(CO_5)\}_6^{2-}$	–	–	508	1848	–	–	[51]
$Sn_6\{(Mo(CO_5)\}_6^{2-}$	–	–	561	1780	–	–	
$Sn_6\{(W(CO_5)\}_6^{2-}$	W	–	198	1803	700	–	
$Sn_7\{Ga\text{-}dppCl\}_2$		–	601	–	–	–	[52]
$Sn_{17}\{Ga\text{-}dppCl\}_4$		–	−836, −896	–	–	–	
$Sn_5\{SnC_6H_3\text{-}2,6\text{-}(C_6H_3\text{-}2,6\text{-}i\text{-}Pr_2)_2\}_2$		–	419, 529	–	–	–	[53]
$Sn_8(2,6\text{-}Mes_2C_6H_3)_4$		–	751, 483	–	–	–	[54]

[a]All $\delta^{119}Sn$ are relative to Me_4Sn. [b,c]The data were taken from [55] and [34] in DMF and NH_3, respectively. See Fig. 5 for atom labeling scheme

3.2 *The $E_9M(CO)_3^{4-}$ Series Where M = Cr, Mo, W; E = Sn, Pb*

This class of complexes is the largest among the Zintl ions. The simplest complexes are $M(CO)_3$ derivatives of the E_9^{4-} ions where M = Cr, Mo, W and E = Sn and Pb [34, 55, 58–60].

These $E_9M(CO)_3^{4-}$ ions are the only E_9 clusters with diminished fluxionality. The structures have virtual C_{4v} symmetry (ignoring the carbonyl ligands) and are quite similar to the *nido*-E_9^{4-} parent ions (Fig. 5). Schrobilgen reported the NMR data for the $M(CO)_3Sn_9^{4-}$ ions concurrently with our group [34, 55]. Each cluster

Table 4 NMR data for the transition metal derivatives of Lead Zintl Ions[a]

A						
Cluster ion	Solvent	Pb 6–9	Pb 2–5	Pb 1	*J* $^{207}Pb–^{207}Pb$	Ref
$Pb_9Mo(CO)_3^{4-}$	en	−1869	−3100	–	–	[34]
	NH_3	−1934	−3450	27	660, 4630	
B	Metal	δ X	$\delta^{207}Pb$	*J* $^{207}Pb–^{195}Pt$		
$NiPb_{12}^{2-}$	–	–	1167	–	–	[56]
$PdPb_{12}^{2-}$	–	–	1520	–	–	
$PtPb_{12}^{2-}$	Pt	−4527	1780	3440	–	
$NiPb_{10}^{2-}$	–	–	−996	–	–	[57]

[a] $\delta^{207}Pb$ chemical shifts are relative to Me_4Pb

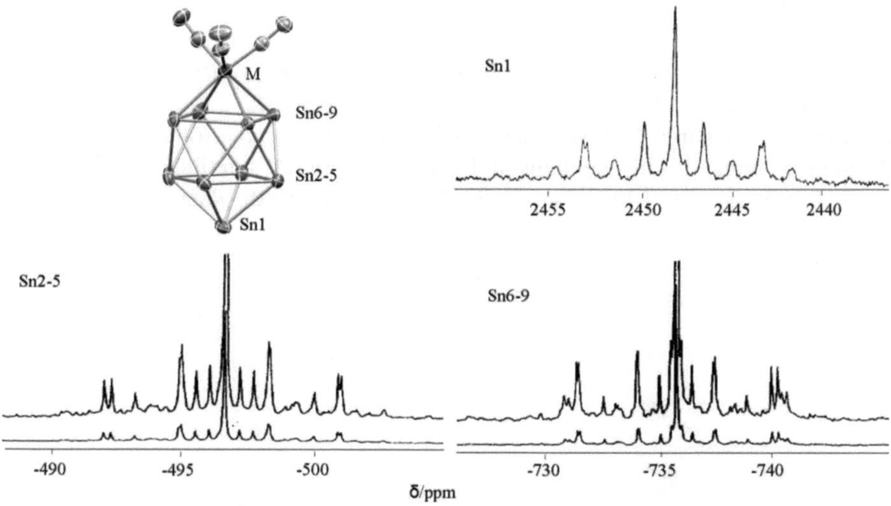

Fig. 5 Structure of η^4-$Sn_9M(CO)_3^{4-}$ ions (M = Cr, Mo, W) and ^{119}Sn NMR of $Sn_9W(CO_3)^{4-}$ anion (expanded spectra on the top). Reprinted with permission [34]. Copyright 2002 American Chemical Society

shows three ^{119}Sn resonances in a 4:4:1 ratio, which is consistent with solid-state structures (Table 3; Fig. 5). The resonances are mutually coupled giving rise to second-order satellite patterns that have been deconvoluted through simulation. The resonances are solvent-dependent and two of the three resonances appear in the range −180 to −736 ppm, which is the "normal range" for Zintl clusters. The resonances for the four-bonded Sn1 atoms appear far downfield in the range 1,988–2,493 ppm and are the most downfield $\delta^{119}Sn$ shifts among the Zintl clusters. The coupling constants are larger than highly fluxional systems with $J(^{119}Sn–^{117}Sn)$ values in the range 322–1,048 Hz with the largest values being those involving Sn1.

The $Pb_9Mo(CO)_3^{4-}$ also shows three ^{207}Pb NMR resonances in a 4:4:1 ratio with the four-bonded Pb1 being the most downfield at 27 ppm. The ^{95}Mo NMR spectrum

(^{95}Mo, 15.8%, I = 5/2) of the compound shows a resonance at δ^{95}Mo = −2,003 ppm with a single J(^{207}Pb–^{95}Mo) coupling of 75 Hz [34].

In addition to the C_{4v} structure of the $M(CO)_3E_9^{4-}$ ions shown in Fig. 5, a second isomer has been characterized for both the tin and lead clusters. $Sn_9W(CO)_3^{4-}$ and $Pb_9Mo(CO)_3^{4-}$ have both been isolated in a C_s structure in which the $M(CO)_3$ fragment occupies a 5-coordinate waist vertex site instead of the 4-coordinate capping site shown in Fig. 5 [55, 60]. Both of these isomers give the same NMR spectra, which is consistent with the higher symmetry C_{4v} structures. We interpret these data to indicate that there is local exchange between the 4-coordinate and 5-coordinate sites, but this interpretation is controversial [34, 55].

3.3 *The $Pt@Sn_9Pt(PPh_3)^{2-}$ Ion*

The $Pt@Sn_9Pt(PPh_3)^{2-}$ ion has quite unusual dynamic properties displaying a "static" Pt–Pt–PPh_3 unit inserted into a highly dynamic Sn_9^{2-} cage (Fig. 6) [49]. The room temperature ^{31}P NMR spectrum is consistent with the solid-state structure showing a single resonance flanked by two sets of Pt satellites: ^{1}J(^{195}Pt–^{31}P) = 4,800 Hz and ^{2}J(^{195}Pt–^{31}P) = 305 Hz. The ^{31}P NMR data show that the

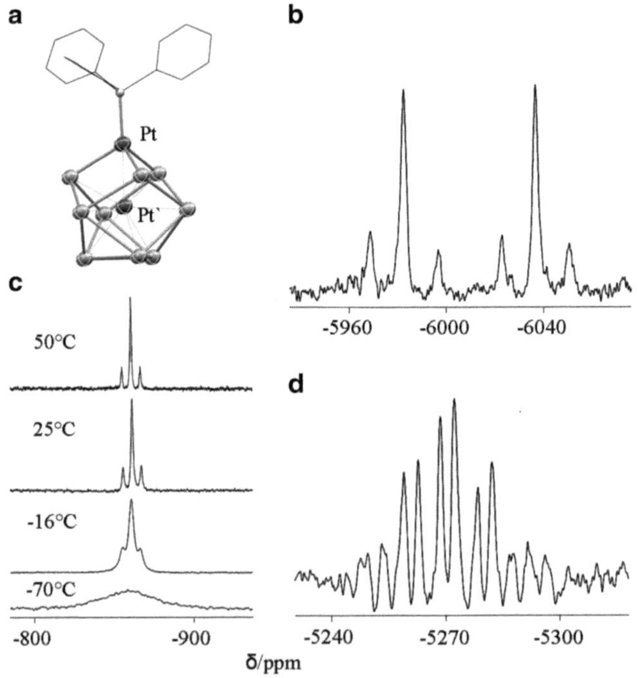

Fig. 6 (**a**) Structure, (**b**, **d**) ^{195}Pt NMR, (**c**) variable temperature ^{119}Sn NMR of the $Pt@Sn_9Pt(PPh_3)^{2-}$ cluster anion. Reprinted with permission [49]. Copyright 2002 American Chemical Society

Pt–Pt–PPh_3 unit remains intact and the two Pt atoms are not in exchange on the NMR time scale (two distinct sets of satellites). In contrast, the ^{119}Sn spectrum of the Pt@$Sn_9Pt(PPh_3)^{2-}$ shows a single ^{119}Sn resonance with a lone set of ^{195}Pt satellites J (^{195}Pt–^{119}Sn) = 1,690 Hz due to coupling only to the interstitial Pt atom (Fig. 6). Coupling to the vertex Pt atom is not observed nor are any ^{117}Sn satellites. Upon cooling, the resonance broadens into the baseline but does not re-emerge at −72°C (DMF/tol, 149 MHz). The ^{195}Pt NMR data show two resonances at δ^{195}Pt = −5,270 and −6,010 ppm arising from the interstitial and vertex Pt atoms, respectively (Fig. 6). Both resonances show Pt satellites due to the one-bond Pt–Pt coupling of 2,472 Hz. The peaks are easily assigned from the ^{31}P couplings of 305 and 4,800 Hz associated with the interstitial (δ^{195}Pt = −5,270 ppm) and vertex (δ^{195}Pt = −6,010 ppm) Pt atoms, respectively. Only the interstitial Pt resonance shows tin couplings J(^{195}Pt–^{119}Sn) = 1,690 Hz which is consistent with the single set of ^{195}Pt satellites on the ^{119}Sn resonance. The Sn–Pt coupling to the vertex Pt atom is too small to resolve.

The dynamic behavior involves rapid exchange of all nine Sn atoms about a static Pt–Pt–PPh_3 fragment. The lack of Pt–Sn coupling to the vertex Pt atom (or its small value) is due to the averaging of the expected large one-bond couplings and small two- and three-bond couplings (~0 Hz) predicted from the solid-state structure. Although Sn–Pt coupling to the vertex Pt atom is not detected, it is clear that the exchange is intramolecular. These data show that centered Pt atoms show very large couplings to the Sn clusters, whereas Pt–Sn coupling to vertex Pt atoms can go undetected in highly dynamic clusters.

3.4 The $E_9PtL_n{}^{q-}$ Complexes Where E = Sn, Pb; Rudolph's Compounds

The first transition metal derivatives of a Zintl ion was prepared by Teixidor et al. in 1983 in reactions between $Pt(PPh_4)_4$ and en solutions of the $E_9{}^{4-}$ (E = Sn, Pb) [25, 26]. Despite being the first examples in this important class of clusters, the complexes have yet to be isolated and their structures and compositions remain unknown. The authors propose that complexes have a $(PPh_3)_2PtSn_9{}^{4-}$ stoichiometry and a *nido*-type structure. Based on comparisons with NMR parameters from the past 30 years and the stoichiometry of the reactions described by Teixidor et al., we believe that the Rudolph compounds are most likely 22-electron *closo*-Pt@E_9Pt $(PPh_3)^{4-}$ complexes. Our rationale is given below.

The reactions between $Pt(PPh_4)_4$ and en solutions of the $Sn_9{}^{4-}$ reproducibly give a single ^{119}Sn NMR resonance at −736 ppm that has large Pt satellites, J(^{195}Pt–^{119}Sn) = 1,554 Hz, due to coupling to *one* Pt atom, and diminished J(^{119}Sn–^{117}Sn) couplings of 79 Hz (Fig. 7) as described by Teixidor et al. The ^{31}P NMR spectrum shows a downfield phosphine peak at 147 ppm with a J(^{195}Pt–^{31}P) coupling of 5,220 Hz due to coupling to *one* Pt atom. These data are the signature of Rudolph's compound. Addition of 2,2,2-crypt to this solution results in the

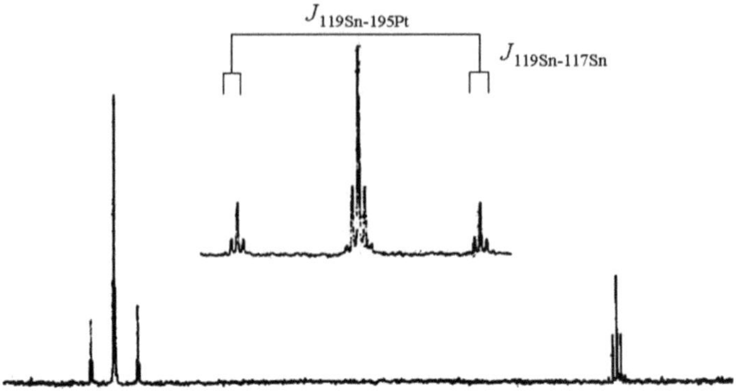

Fig. 7 ^{119}Sn NMR of Rudolph's $K_xL_2Pt_ySn_9$ (left and inset) and Sn_9^{4-} (right) clusters. Reprinted with permission [25]. Copyright 1983 American Chemical Society

disappearance of Rudolph's compound, the evolution of H_2 gas, and the formation of the $Pt@Sn_9Pt(PPh_3)^{2-}$ ion described above [49]. The latter complex clearly has two Pt atoms – one centered in the Sn_9 cage and the other occupying a vertex site – but $J(^{195}Pt–^{119}Sn)$ coupling is only observed between the centered Pt and Sn_9 cluster as is of similar magnitude ($J(^{195}Pt–^{119}Sn) = 1{,}690$ Hz) as Rudolph's complex. The coupling between the capping $Pt(PPh_3)$ and the Sn_9 is not observed (<50 Hz) due to the dynamic properties of the Sn_9 cage, again similar to Rudolph's complex [25]. The ^{31}P NMR resonance for $Pt@Sn_9Pt(PPh_3)^{2-}$ ion shows coupling to 2 Pt atoms; one large coupling of 4,800 Hz, which is similar to the 5,220 Hz coupling observed for Rudolph's compound. Titration experiments in our lab and in Rudolph's paper also suggest that the stoichiometry of the reaction is 2Pt to $1Sn_9^{4-}$. For this reason, it is possible that Rudolph's complex is the $Pt@Sn_9Pt(PPh_3)^{4-}$ ion, which can undergo a two-electron oxidation to give the $Pt@Sn_9Pt(PPh_3)^{2-}$ ion [49] as outlined in (1) below.

$$\begin{aligned} &Pt@Sn_9Pt(PPh_3)^{4-} + 2H_2NCH_2CH_2NH_2 \\ &\quad \rightarrow Pt@Sn_9Pt(PPh_3)^{2-} + H_2 + 2H_2NCH_2CH_2NH^{-1} \end{aligned} \qquad (1)$$

The lead complex described by Rudolph [25, 26] also has a single ^{207}Pb NMR resonance at $-2{,}988$ ppm that is flanked by Pt satellites $J(^{207}Pb–^{195}Pt) = 4{,}122$ Hz. The chemical shift is in the correct range for an $L_nMPb_9^{4-}$ type cluster (see below), but the large $^{207}Pb–^{195}Pt$ coupling constant is suggestive of a centered Pt atom. For example, the Pt-centered $Pt@Pb_{12}^{2-}$ ion shows [56] a similar coupling of $J(^{207}Pb–^{195}Pt) = 3{,}440$ Hz (see below). As described above, the $Pt@Sn_9Pt(PPh_3)^{2-}$ ion shows only coupling between the centered Pt and Sn_9 cluster; the coupling between the capping $Pt(PPh_3)$ and the Sn_9 is not observed (<50 Hz). Addition of 2,2,2-cryptand or other alkali ion sequestering agents to these solutions results in the formation of the $Pt@Pb_{12}^{2-}$ cluster described below. For the reasons described above, we propose that Rudolph's Pb cluster also has a 2Pt:$1Pb_9^{4-}$ ratio and has the formula $Pt@Pb_9Pt(PPh_3)^{4-}$. The two sets of ^{195}Pt satellites present on the ^{31}P NMR resonance are very similar to those observed for $Pt@Sn_9Pt(PPh_3)^{2-}$ and are supportive of this assignment.

3.5 $Cu@Sn_9^{3-}$ and $Sn_9Ir(COD)^{3-}$

These two clusters are both diamagnetic derivatives of formally M^{+1} metals, but they have very different structures and spectroscopic properties. The $Cu@Sn_9^{3-}$ ion has a centered Cu atom in a D_{3h} – type Sn_9 cluster; however, the solid-state structure has only approximate C_{2v} symmetry [44]. Regardless, the complex is fluxional in solution giving a single ^{119}Sn resonance at −1,440 ppm with $J(^{119}Sn–^{117}Sn) = 85$ Hz. The remarkable feature of the spectrum is the well-resolved coupling to the quadrupolar Cu atom (^{63}Cu, 69.1% abund., I = 3/2; ^{65}Cu, 30.9% abund., I = 3/2) with a $J(^{117/119}Sn–^{63/65}Cu)$ coupling constant of 286 Hz (Fig. 8). The ^{63}Cu NMR spectrum is also unusually sharp (Fig. 8), showing a $^{117/119}Sn$ satellite pattern indicative of a highly fluxional, 9-atom cluster. The line intensities of the ^{63}Cu signal are in agreement with the expected intensities; 0.006:0.04:0.22:0.63:1:0.63:0.22:0.04. Suppression of quadrupolar broadening only occurs in high symmetry cubic symmetries, which is generated through the dynamic exchange of the Sn_9 coordination sphere.

The $Sn_9Ir(COD)^{3-}$ ion formally has an Ir^{+1} ion and is a diamagnetic 10-vertex, 22-electron *closo*-cluster [45, 61]. In contrast to the isoelectronic $E_9M(CO)_3^{4-}$ ions described above, the Ir cluster shows a single ^{119}Sn resonance at −1,149 ppm indicative of fast global exchange of all nine Sn atoms at room temperature. The ^{117}Sn satellites $J(^{119}Sn–^{117}Sn = 139$ Hz) are less than the parent Sn_9^{4-} precursor [45].

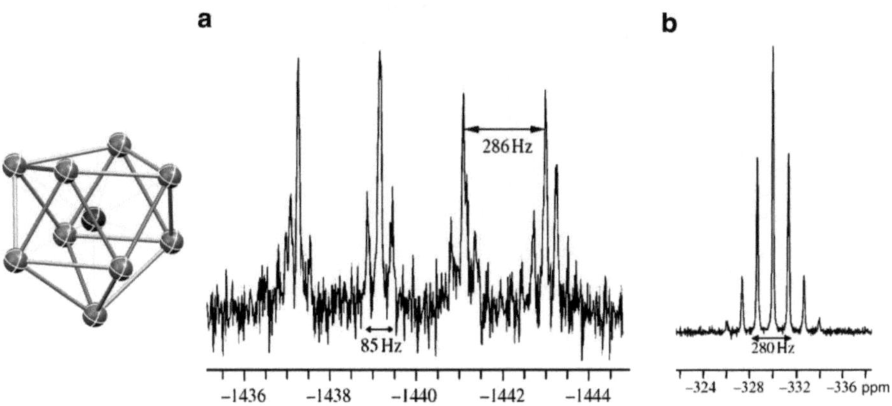

Fig. 8 Structure, ^{119}Sn (**a**) and ^{63}Cu NMR spectra for the $Cu@Sn_9^{3-}$ cluster. Reprinted with permission [44]. Copyright Wiley-VCH Verlag GmbH & Co. KGaA. Reproduced with permission

3.6 The $M@Sn_9H^{3-}$ ions (M = Ni, Pd, Pt)

The similarity of the NMR spectroscopic properties for the $M@Sn_9H^{3-}$ ions (M = Ni, Pd, Pt) suggests that they are isostructural, although only the $Ni@Sn_9H^{3-}$ and $Pd@Sn_9H^{3-}$ ions have been structurally characterized to date (Fig. 9) [48, 50]. The dynamic properties are most clearly evident from the $Pt@Sn_9H^{3-}$ cluster, which are highlighted below. The ^{119}Sn NMR data for all three complexes show single, time-averaged resonances between −368 and −837 ppm with well-defined ^{117}Sn satellites J(^{119}Sn–^{117}Sn ~80 Hz). The peaks are split into doublets due to coupling to the hydride ligands J(^{119}Sn–^{1}H ≈ 46 Hz). The ^{1}H NMR signals for the hydride ligands appear between −28 and −22.3 ppm with satellite intensities and patterns consistent with the calculated distributions of any heteroatom coupled to nine equivalent Sn nuclei (see Sect. 3.5). The ^{119}Sn resonance for the $Pt@Sn_9H^{3-}$ complex also displays large Pt–Sn coupling J(^{195}Pt–^{119}Sn = 1,540 Hz, 33% intensity, Fig. 9) that is indicative of an interstitial Pt atom. The most unusual feature of this series involves the fluxionality of the hydride ligands. The high field ^{1}H NMR chemical shifts are suggestive of transition metal hydride complexes, but the J(^{195}Pt–^{1}H) coupling constant of 32 Hz in $Pt@Sn_9H^{3-}$ (Fig. 9) is much smaller than those of typical Pt–H species. For example, the Pt–H coupling constant in the P_7PtH

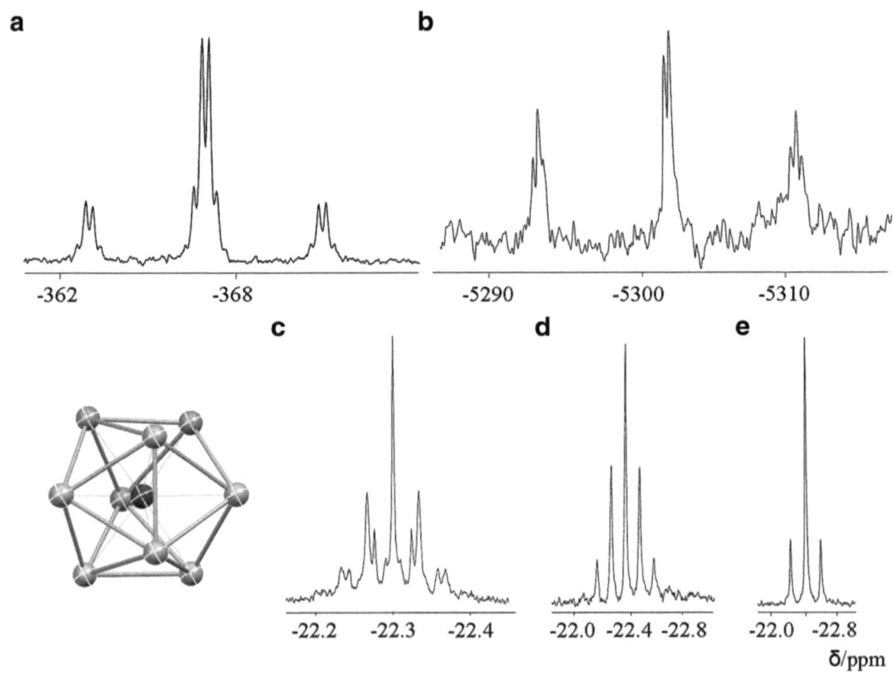

Fig. 9 Structure of $M@Sn_9H^{3-}$ clusters (M = Ni, Pd) and (**a**) the ^{119}Sn NMR , (**b**) the ^{195}Pt NMR, (**c**) the ^{1}H NMR, (**d**) the $^{1}H\{^{195}Pt\}$ NMR, (**e**) the $^{1}H\{^{119/117}Sn\}$ NMR spectra for $Pt@Sn_9H^{3-}$ cluster. Reprinted with permission [48, 50]. Copyright 2007 American Chemical Society

$(PPh_3)^{2-}$ complex is 1,080 Hz [62]. The observed ~46 Hz $^{119}Sn–^{1}H$ coupling constants in all three clusters are also much less than the typical $^{119}Sn–^{1}H$ values, such as in Bu_2SnIH where $^{1}J(^{119}Sn–^{1}H) = 2{,}060$ Hz [63]. The $[Sn(Ar)H]_2$ dimers that contain bridging hydrides show somewhat smaller $^{1}J(^{119}Sn–^{1}H)$ values of ~89 Hz [64]. The small coupling constants suggest that the H atom is rapidly scrambling around the outside surface of the cluster between the Pt and Sn atoms. The single Sn resonance suggests that the Sn atoms in the Sn_9 cages are also in rapid exchange. Both processes will generate time-averaged values significantly less than expected static, single bond couplings. Although the scrambling is fast on the NMR time scale, the process must be intramolecular as coupling is maintained in the fast exchange limit. By contrast, the hydride ligands in the closely related $E_7PtH(PPh_3)^{2-}$ ions (E = P, As) apparently do not scramble across the cluster surfaces nor do they form semibridging sites [62]. However, the site of hydrogen attachment can vary between the main group cage and the transition metal on closely related clusters [e.g., $P_7PtH(PPh_3)^{2-}$ with Pt–H and $P_7NiH(PPh_3)^{2-}$ with a P–H] [62].

3.7 The Paramagnetic $Ni@Sn_9Ni(CO)^{3-}$ Ion

The dynamic properties of paramagnetic clusters are difficult to assess experimentally. The EPR spectra of the paramagetic $E_9{}^{3-}$ ions are not particularly informative due to broad, weak signals and lack of hyperfine interactions. However, the EPR spectrum of $Ni@Sn_9Ni(CO)^{3-}$ ion where S = ½ (Fig. 10) shows well-defined hyperfine interactions of 4.8 G due to interactions with the nine tin atoms [49]. Equal contributions from all nine Sn atoms are consistent with a dynamic exchange but certainly not proof of fluxionality. While it is possible that the unpaired electron simply couples equally to all Sn atoms in the static structure, it is more likely that the cluster is dynamic. All of the diamagnetic tin clusters containing interstitial metal atoms have been shown to be highly fluxional.

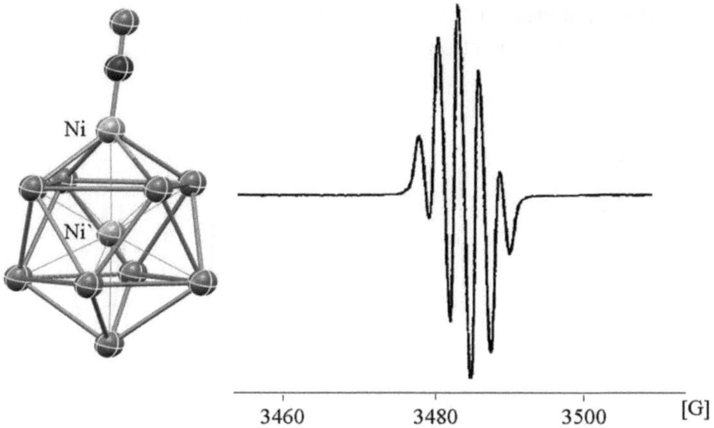

Fig. 10 Structure and EPR spectrum of the $Ni@Sn_9Ni(CO)^{3-}$. Reprinted with permission from [49]. Copyright 2002 American Chemical Society

3.8 The $Ni@Pb_{10}^{2-}$ Ion

The bicapped square pyramidal D_{4d} structure of $Ni@Pb_{10}^{2-}$ has two chemically distinct lead environments, but its ^{207}Pb NMR spectrum shows a single, somewhat broad resonance at −996 ppm (25°C, 104.7 MHz) as a result of fast exchange [57]. As the $Ni@Pb_{10}^{2-}$ sample is cooled to −45°C, the signal broadens significantly due to slowing of intramolecular exchange on the NMR time scale, but the limiting spectrum has not been achieved.

3.9 The $M@Pb_{12}^{2-}$ Ions Where M = Pt, Pd, Ni

The $M@Pb_{12}^{2-}$ ions have nearly perfect I_h point symmetries with 12 equivalent Pb atoms in icosahedral geometries [56]. Their dynamic behavior remains unknown due to the high symmetry, but NMR studies of the isostructural, isoelectronic $Ir@Sn_{12}^{3-}$ ion may provide insight into possible exchange in this class of clusters. The ^{207}Pb NMR chemical shifts of $M@Pb_{12}^{2-}$ ions are significantly downfield relative to other known lead clusters (Table 4) [65]. Typical lead compounds have chemical shifts in the range of 0 to −4,500 ppm, and most polyanions are in the upfield portion of this range (e.g., Pb_9^{4-} at −4,098 ppm). Chen, Schleyer and King have attributed these shifts to aromatic character associated with the high symmetry I_h structures [66–68]. Support for the symmetry effects can be found in the comparison between the $Ni@Pb_{10}^{2-}$ and $Ni@Pb_{12}^{2-}$, which have similar structures, charges, and nuclearities, but their ^{207}Pb NMR chemical shifts differ by more than 2,000 ppm.

The Pd and Ni complexes give rise to simple, single-line resonances; however, the ^{207}Pb NMR spectrum for $Pt@Pb_{12}^{2-}$ shows a singlet flanked by Pt satellites $^{1}J(^{207}Pb–^{195}Pt \approx 3{,}440)$ indicating strong coupling to the central Pt atom (Fig. 11). The observed satellite patterns on the ^{207}Pb and ^{195}Pt signals are in agreement with the calculated satellite patterns of 2.65:1:2.65 and 0.4:0.8:1:0.8:0.4, respectively. The corresponding ^{195}Pt NMR spectrum shows a single resonance at −4,527 ppm with

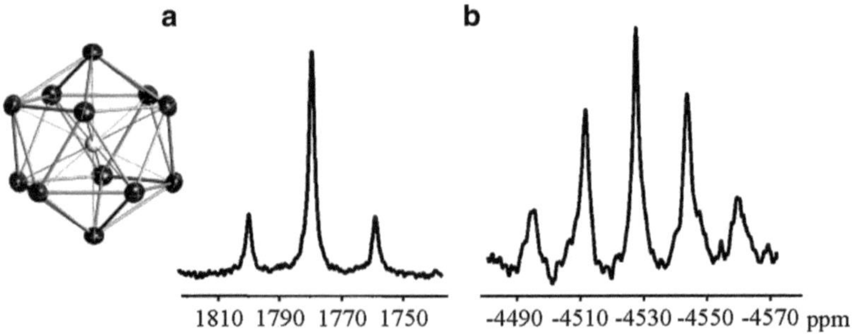

Fig. 11 Structure and (**a**) the ^{207}Pb, and (**b**) the ^{195}Pt NMR spectra for the $Pt@Pb_{12}^{2-}$. Reprinted with permission [56]. Copyright 2006 American Chemical Society

a satellite pattern arising from coupling to 12 equivalent Pb. Both spectra are consistent with the nuclearity of the cluster and its solid-state structure.

3.10 The $Pt_2@Sn_{17}{}^{4-}$ and $Ni_2@Sn_{17}{}^{4-}$ Ions

The dynamic properties of these two complexes are perhaps the most surprising and interesting of the transition metal Zintl ion complexes. The clusters have markedly different structures but are both prolate in nature with several different Sn environments. The tin atoms at opposite ends of both structures are separated by nearly 9 Å, but all 17 atoms are in fast intramolecular exchange on the NMR time scale.

The room temperature ^{119}Sn spectrum of the $Pt_2@Sn_{17}{}^{4-}$ complex shows a single ^{119}Sn resonance at −742.3 ppm flanked by one set of ^{195}Pt satellites with $J(^{195}Pt–^{119}Sn) = 774$ Hz due to the coupling to the two interstitial Pt atoms [48]. The exchange process is intramolecular (no loss of coupling) and remains rapid on the NMR time scale at −60°C. The intensities of the ^{195}Pt satellites indicate equal coupling to the two equivalent Pt atoms of the $Pt_2@Sn_{17}{}^{4-}$cluster. The ^{119}Sn resonance has ^{117}Sn satellites with $J(^{119}Sn–^{117}Sn) = 170$ Hz and intensities consistent with a 17-atom tin cluster. The ^{195}Pt NMR spectrum also shows a single resonance, centered at −5,713 ppm, with a satellite pattern due to coupling to all 17 tin atoms. The coupling constant $^1J(^{195}Pt–^{119/117}Sn) \approx 780$ Hz equals that observed in the ^{119}Sn NMR spectrum.

This coupling constant is approximately half of the $^{195}Pt–^{119}Sn$ couplings observed for the centered Pt atoms in the nine-atom clusters $Pt@Sn_9Pt(PPh_3)^{2-}$, $Pt@Sn_9H^{3-}$, and Rudolph's compound (Table 3). In the latter clusters, each Pt maintains direct bonding contact with each Sn atom during dynamic exchange. In contrast, direct bonding between a given Pt–Sn pair is only present 50% of the time in the $Pt_2@Sn_{17}{}^{4-}$ complex, thereby reducing the coupling constant by half. The intensities of the Pt satellites and the magnitude of the $J(^{195}Pt–^{119/117}Sn)$ coupling constants are potent indicators of both cluster nuclearity and the mechanism of dynamic exchange.

The low temperature limiting ^{119}Sn NMR spectrum of $Ni_2Sn_{17}{}^{4-}$ at −64°C (Fig. 12) contains four ^{119}Sn resonances in the range −1,713 to 228 ppm, which are consistent with the D_{2d} structure with four different Sn environments [46]. When the temperature is increased to 60°C, the four peaks become time averaged and a single resonance is observed at −1,167 ppm. Based on the weighted average of the low temperature limiting spectrum, the expected time average peak for total exchange is −1,120 ppm. At room temperature, the peak due to Sn5 remains distinct, whereas the other three show a time-averaged resonance. These data suggest that an intramolecular exchange process is occurring at room temperature that does not involve the central Sn5 atom. Because the 8-coordinate Sn5 atom is more akin to solid-state Sn coordination environments, this process serves as a model for different exchange rates between surface and bulk atoms in NP systems. Fischer and coworkers reported a gallium stabilized Sn_{17} cluster $[(ddp)ClGa]_4Sn_{17}$ that has a similar

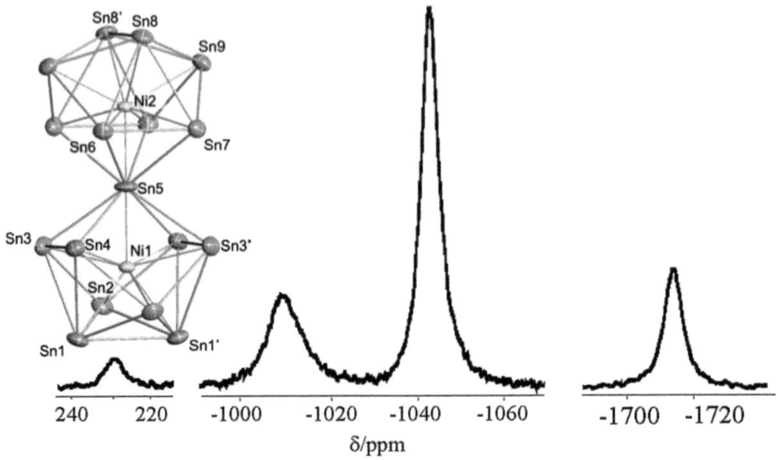

Fig. 12 Structure and ^{119}Sn NMR spectra of $Ni_2@Sn_{17}^{4-}$. Reprinted with permission [46]. Copyright 2006 American Chemical Society

Sn_{17}^{4-} framework [ddp = HC(CMeNC$_6$H$_3$-2,6-^{i}Pr$_2$)$_2$] [52]. This compound also appears to be highly fluxional giving only two broad resonances in the solid state.

The site exchange in the $Pt_2@Sn_{17}^{4-}$ ion is apparently faster than that in the isoelectronic $Ni_2@Sn_{17}^{4-}$ cluster. As such, the limiting ^{119}Sn spectrum can be obtained for $Ni_2@Sn_{17}^{4-}$ but not for $Pt_2@Sn_{17}^{4-}$ under identical conditions. The $Pt_2@Sn_{17}^{4-}$ structure may be a transition state in the $Ni_2@Sn_{17}^{4-}$ exchange process.

3.11 *The $Pd_2@Sn_{18}^{4-}$ Ion*

The largest global dynamic exchange observed among Zintl ions in general involves the 18-atom exchange in the $Pd_2@Sn_{18}^{4-}$ ion [47, 69]. This cluster gives rise to a single resonance at $\delta^{119}Sn = -734$ ppm. Exchange remains rapid on the NMR time scale down to −60°C in dmf solution. Unlike the $Pt_2@Sn_{17}^{4-}$ cluster, the ^{117}Sn–^{119}Sn coupling in the $Pd_2@Sn_{18}^{4-}$ ion was not detected (<70 Hz), suggesting that the exchange mechanisms may be substantially different.

4 Exo-Bonded Substituents and Other Main Group Derivatives of the E_9^{4-} Ions

4.1 *The Tl–Sn^{n-} and Bi–Sn^{n-} Anions*

Most of the bare main group derivatives of the E_9^{4-} ions were prepared several years ago by Rudolph and Corbett, whereas the exo-bonded compounds were

Table 5 Main group derivatives and exo-bonded derivatives of Group 14 Zintl ions[a]

Cluster ion	Cation	$\delta^{119}Sn$	*J*			Ref
			$^{119}Sn–^{117}Sn$	$^{119}Sn–^{207}Pb$	$^{119}Sn–^{205}Tl$	
Sn_8Tl^{5-} [b]	Na	−1169	412	–	800	[27]
	K	−1137	456	–	646	
Sn_7PbTl^{5-}	Na	−1218	418	335	791	[23]
	K	−1183	–	–	665	
$Sn_6Pb_2Tl^{5-}$	Na	−1266	444	366	781	
	K	−1227	–	–	644	
$Sn_5Pb_3Tl^{5-}$	K	−1270	–	–	593	
$Sn_4Pb_4Tl^{5-}$	K	−1317	–	–	519	
$Sn_2Bi_2^{2-}$	Li	−1674	1606	–	–	
	Na	−1574	1638	–	–	
	K	−1575	1638	–	–	
	Sn_9R^{3-}	$\delta^{119}Sn$	$^{119}Sn–^{117}Sn$	$^{119}Sn–^{119}Sn$	$^{119}Sn–^{117}Sn$	
Alkyl substituted	$Sn_9CCH_3^{3-}$	−1417, 255	121	1876	1795	Kocak and Eichhorn, unpublished results
	$Sn_9CH(CH_3)_2^{3-}$	−1413, 170	115	1876	1793	[70]
–SnR_3 substituted	$Sn_9Sn(C_6H_{11})_3^{3-}$	−1172, 155	295	1331	1272	
	[c]$Sn_9Sn(C_4H_9)_3^{3-}$	−1119, 62	278	1330		Kocak and Eichhorn, unpublished results

[a]All $\delta^{119}Sn$ are relative to Me_4Sn
[b]Data were collected in NH_3 −34°C
[c]Data collected in pyridine

prepared more recently. A summary of the NMR data for these compounds can be found in Table 5.

Both Rudolph and Corbett investigated ethylenediamine extracts of A–Tl–Sn melts (A = Na, K). Burns and Corbett crystallized a 50:50 mixture of $TlSn_9^{3-}$ and $TlSn_8^{3-}$ that was characterized by single-crystal X-ray diffraction [71]. Rudolph and Wilson observed only one diamagnetic cluster, which they proposed to be $TlSn_8^{5-}$ [27]. Like Sn_9^{4-}, the $TlSn_8$ cluster is highly fluxional showing a single time-averaged ^{119}Sn resonance at −1,169 ppm in (liq-NH_3, −34°C) with distinct coupling to thallium (^{203}Tl, 29.5% abund., I = ½; ^{205}Tl, 70.48% abund., I = ½) with $J(^{119}Sn–^{205,203}Tl)$ = 800 Hz and pronounced tin satellites $J(^{119}Sn–^{117}Sn)$ = 410 Hz (Fig. 13). The experimental line intensities are in agreement with the calculated two-line satellite pattern of the 0.03:0.27:1:0.27:0.03. Wilson et al. also reported data for the mixed Sn/Pb/Tl clusters of proposed composition $TlSn_{8-x}Pb_x^{5-}$ (Fig. 13) [28]. The relationships between these clusters remain unclear and further studies are warranted.

The tetrahedral $Sn_2Bi_2^{2-}$ ion has been structurally characterized by Corbett [72] and studied in solution by Rudolph [23]. The ^{119}Sn NMR resonance at −1,574 ppm shows coupling to one other Sn atom with a large, one-bond $J(^{119}Sn–^{117}Sn)$ = 1,638 Hz. Recent gas phase and theoretical studies on Bi-substituted nine-atom cluster anions suggest that the isoelectronic Sn_9^{4-} analog, $Sn_6Bi_3^{1-}$, should be static in solution [31].

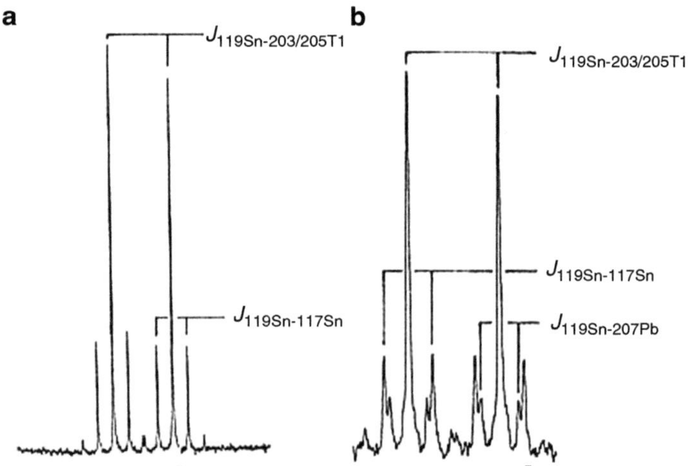

Fig. 13 ^{119}Sn NMR spectra of Sn_8Tl^{5-} (*left*) [23] Sn_8PbTl^{5-} (*right*). Reprinted with permission [23, 28]. Copyright 1986 American Chemical Society

4.2 Exo-Bonded Substituents on Polystannide Clusters

It has been shown that a variety of substituents can be attached to the outside of the group 14 Zintl ion clusters in *exo* positions (i.e., not vertex or interstitial positions) [70, 73–78]. A variety of alkyl, aryl, and main group moieties have been attached to Ge_9 and Sn_9 clusters. The structures of these clusters are similar to some organostannane clusters prepared via different synthetic routes. This burgeoning class of compounds is rapidly developing; however, little is known about the effect of the *exo*-substituents on the dynamic properties of the clusters. Only the RSn_9^{3-} ions, where R = i-Pr, t-Bu, and $SnCy_3$, Sn–*n*–Bu_3, have been studied in detail [70].

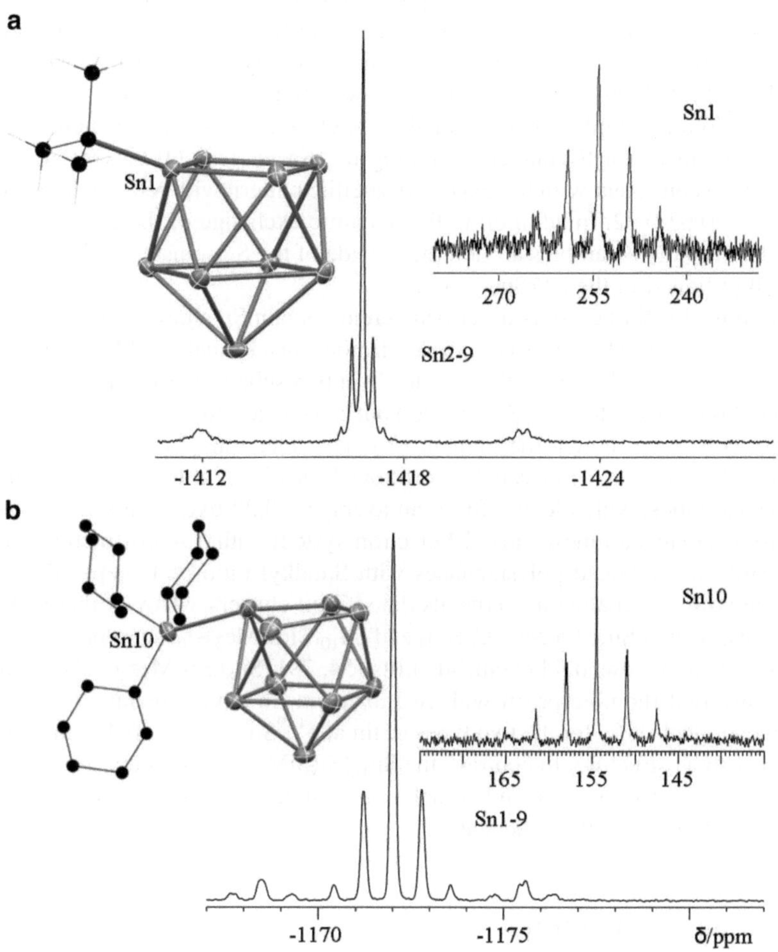

Fig. 14 Structures and ^{119}Sn NMR spectra for (**a**) the Sn_9-*t*-Bu^{3-} and (**b**) the $Sn_9SnCy_3^{3-}$. Reprinted with permission [70, 74], Kocak and Eichhorn unpublished results

All Sn_9R^{3-} clusters are highly fluxional, showing rapid exchange of the Sn atoms of the cluster's vertices (Fig. 14). The ^{119}Sn chemical shifts of the exchanging cluster vertex atoms appear in the range −1,119 to −1,413 ppm (Table 5). The upfield signal for all the clusters shows coupling to a single Sn atom with an approximately 16% intensity. Although the data appear similar, the mechanisms of exchange are quite different [70]. The Sn–C bond of the i-$PrSn_9^{3-}$ ion is non-labile, which precludes chemical exchange of the Sn(1) vertex with the other eight vertex Sn atoms. As such, the chemical shift of Sn1 remains unique ($\delta^{119}Sn = 119$ ppm) and shows large $J(^{119}Sn–^{119}Sn)$ and $J(^{119}Sn–^{117}Sn)$ couplings of 1,876 and 1,793 Hz, respectively. In agreement with the calculated satellite pattern, the Sn1 signal shows satellites with intensities of 0.18:0.60:1:0.60:0.18. The small 115 Hz $^{119}Sn–^{117}Sn$ satellites on the −1,413 ppm central resonance suggest that the exchange mechanism goes through a more open transition state due to the presence of the alkyl group.

The dynamic exchange of the Sn_9 core of the $SnCy_3Sn_9^{3-}$ ion appears to go through the traditional "diamond–square–diamond" exchange mechanism as evidenced by the typical 295 Hz $^{119}Sn–^{117}Sn$ satellites on the −1,172 ppm central resonance. While coupling pattern of the upfield signal is in agreement with the expected 9-equivalent Sn-nucleus coupling to 1Sn nucleus (16%), the downfield signal is also consistent with the expected satellite pattern with relative intensites of 0.22:0.63:1:0.63:0.22. In addition to the dynamic exchange of the core, the Cy_3Sn substituent rapidly scrambles around the outside of the Sn_9 cluster in sharp contrast to the alkyl ligand in the i-$PrSn_9^{3-}$ ion.

The group 14 Zintl clusters described herein contain Sn atoms in negative oxidation states. For example, Sn_9^{4-} comprises tin atoms in a formal −0.44 oxidation state. Functionalizing the clusters with *exo*-substituents results in the formation of highly covalent Sn–R bonds (R = C, Sn, Sb, etc.) but can be viewed as a formal oxidation of the Zintl core even though the number of cluster electrons remains the same. For example, the i-$PrSn_9^{3-}$ ion can be partitioned as i-Pr^{-1} and Sn_9^{2-} according to electronegativities, which leaves Sn in an average −0.22 oxidation state. However, the cluster remains a nine vertex, 22-electron system with a *nido*-structure type. In this regard, the alkylated polystannanes with Sn:alkyl ratios >1 prepared from Sn halides are directly analogous to the alkylated Zintl clusters. Many of these alkyl tin complexes, such as $Sn_8(2,6\text{-}Mes_2C_6H_3)_4$ and $Sn_{10}[Si(SiMe_3)_3]_6$, are typified by bulky ligands and show downfield chemical shifts [54, 79]. $Sn_8(2,6\text{-}Mes_2C_6H_3)_4$ contains an Sn_8 distorted rhombic prism with two different tin environments. It shows two distinct chemical shifts for the two types of tin at $\delta^{119}Sn = 752$ and 483 ppm, which signifies a static structure. In contrast, the $Sn_{10}[Si(SiMe_3)_3]_6$ complex shows a single time-averaged $Si(SiMe_3)_3$ ligand in solution despite the three chemically distinct environments found in the solid state.

5 Summary and Outlook

From the pioneering work of Rudolph (group 14 Zintl ions) and Baudler (polyphosphide Zintl ions) to the modern multidimensional techniques used in current research, multinuclear NMR has played a central role in evaluating structures, properties, and dynamic exchange in Zintl ions and their derivatives. Evaluation of data collected over the last 30 years shows that satellite patterns and coupling constants can provide significant insight into the number of spin-active vertex atoms in the cluster, the positions of metal atoms in a cluster (i.e., interstitial or vertex), and the number of atoms in exchange. For example, when an X nucleus couples to nine equivalent Sn atoms, a characteristic satellite pattern is observed regardless of the identity of X (e.g., ^{195}Pt, $^{63/65}Cu$, ^{1}H, $^{119/117}Sn$). Moreover, the magnitudes of the coupling constants provide insight into mechanisms of dynamic exchange.

The generation of very high symmetry interstitial sites in the icosahedral $E_{12}{}^{2-}$ clusters (E = Pb, Sn) provides unique access to spectroscopic studies inaccessible by other means. The I_h point symmetries should allow for the observation of coupling to highly quadrupolar nuclei due to the removal of the quadrupolar broadening as a result of the high ligand fields. Fässler's observation of coupling between $^{119/117}Sn$ and the quadrupolar nuclei $^{63/65}Cu$ shows that even time-averaged cubic environments can provide access to these effects. Similarly, the ^{59}Co NMR signal for $Co@Ge_{10}{}^{3-}$ at −731 ppm is consistent with this hypothesis [80]. Moreover, the I_h coordination environment of the $E_{12}{}^{2-}$ endohedral sites imparts no crystal field splitting on interstital transition metals. Because the *d*-orbitals degenerate such an environment, unusual electronic states can be imagined, such as high-spin 4*d* and 5*d* transition metal complexes. Expanding this class of $M@E_{12}{}^{2-}$ complexes promises to advance both goals.

Finally, the dynamic properties within this class of clusters are truly remarkable in view of the dimension of exchange across large clusters and the very low energy barriers for these processes. The high nuclearities of the $Pd_2@Sn_{18}{}^{4-}$ and $M_2@Sn_{17}{}^{4-}$ clusters (M = Pt, Ni) lead to high densities of orbital states near the HOMO that approaches the density of states of small metallic nanoparticles. As such, the core fluxionality and substrate mobilities (i.e., H, R, SnR_3) observed in these clusters have particular relevance to small bimetallic NP catalysts and their restructuring properties. For example, hydrogenation chemistry catalyzed by PtSn bimetallics is thought to involve hydrogen spillover to the Pt atoms of the bimetallic. However, the dynamic properties of the $Pt@Sn_9H^{3-}$ ion suggest that hydrogen atoms scramble from Pt to Sn with high frequency. Moreover, rapid surface and whole particle restructuring of bimetallic catalysts, which is now thought to be far more facile than first believed, is entirely consistent with the dynamic properties displayed by the large bimetallic clusters described here. Synthesis and study of ever-larger cluster compounds promise to provide more insight into the transition from large molecular clusters to small metallic particles.

Acknowledgments The authors thank the NSF for support of this work through grant CHE-0944528. We are also very grateful to the graduate students from Maryland who performed many of the experiments outlined in this review; namely, Dr. Scott Charles, Domonique Downing, Donna Gardner, Dr. Banu Kesanli, Dr. Emren Esenturk, Dr. Melanie Moses, and Lingling Wang. We are forever indebted to Dr. Yiu-Fai Lam for insight, technical expertise, and invaluable assistance with the NMR studies conducted at the University of Maryland.

References

1. Sita LR (1995) Structure/property Relationships of Polystannanes. In: Stone FGA, West R (eds) Advances in organometallic chemistry, vol 38. Academic, London, pp 189–243
2. King RB (2005) Polyhedral dynamics and the Jahn-Teller effect. In: Gielen M, Willem R, Wrackmeyer B (eds) Fluxional organometallic and coordination compounds. Wiley, New York, pp 1–37
3. Scharfe S, Fässler TF (2010) Polyhedral nine-atom clusters of tetrel elements and intermetalloid derivatives. Phil Trans Royal Soc A Math Phys Eng Sci 368:1265–1284
4. Sevov SC, Goicoechea JM (2006) Chemistry of deltahedral Zintl ions. Organometallics 25:5678–5692
5. Corbett JD (1985) Polyatomic Zintl anions of the post-transition elements. Chem Rev 85:383–397
6. Somorjai G, Contreras A, Montano M et al (2006) Clusters, surfaces, and catalysis. Proc Nat Acad Sci U S A 103:10577–10583
7. Alayoglu S, Nilekar AU, Mavrikakis M et al (2008) Ru-Pt core-shell nanoparticles for preferential oxidation of carbon monoxide in hydrogen. Nat Mater 7:333–338
8. Greeley J, Mavrikakis M (2004) Alloy catalysts designed from first principles. Nat Mater 3:810–815
9. Fischer FD, Waitz T, Vollath D et al (2008) On the role of surface energy and surface stress in phase-transforming nanoparticles. Prog Mater Sci 53:481–527
10. Tao F, Grass M, Zhang Y, Butcher D et al (2008) Reaction-driven restructuring of Rh-Pd and Pt-Pd core-shell nanoparticles. Science 322:932–934
11. Nørskov JK, Bligaard T, Logadottir A et al (2002) Universality in heterogeneous catalysis. J Catal 209:275–278
12. Wang Y, Toshima N (1997) Preparation of Pd-Pt bimetallic colloids with controllable core/shell structures. J Phys Chem B 101:5301–5306
13. Zhou S, Yin H, Wu Z et al (2008) NiAu alloy nanoparticles for preparing highly active Au/NiO_x CO oxidation catalysts. Chem Phys Chem 9:2475–2479
14. Liu Z, Jackson G, Eichhorn B (2010) PtSn intermetallic, core-shell and alloy nanoparticles as CO-tolerant electrocatalysts for H_2 oxidation. Angew Chem Int Ed 49:3173–3176
15. Haushalter RC, O'Connor CM, Haushalter JP et al (1984) Synthesis of new amorphous metallic spin-glasses Cr_2SnTe_4, Mn_2SnTe_4, Fe_2SnTe_4, Co_2SnTe_4 – solvent induced metal-insulator transformations. Angew Chem Int Ed 23(2):169–170
16. Sun D, Riley AE, Cadby AJ et al (2006) Hexagonal nanoporous germanium through surfactant-driven self-assembly of Zintl clusters. Nature 441:1126–1130
17. Guloy AM, Ramlau R, Tang ZJ et al (2006) A guest-free germanium clathrate. Nature 443:320–323
18. Armatas GS, Kanatzidis MG (2008) High-surface-area mesoporous germanium from oxidative polymerization of the deltahedral $[(Ge_9]^{(4-)}$ cluster: electronic structure modulation with donor and acceptor molecules. Adv Mater 20:546
19. Xu L, Sevov SC (1999) Oxidative coupling of deltahedral $[Ge_9]^{4-}$ Zintl ions. J Am Chem Soc 121:9245–9246
20. Ugrinov A, Sevov SC (2003) $[Ge_9{=}Ge_9{=}Ge_9{=}Ge_9]^{(8-)}$: a linear tetramer of nine-atom germanium clusters, a nanorod. Inorg Chem 42:5789
21. Ugrinov A, Sevov SC (2002) $[Ge_9{=}Ge_9{=}Ge_9]^{(6-)}$: A linear trimer of 27 germanium atoms. J Am Chem Soc 124:10990–10991
22. Corbett JD, Edwards PA (1977) The Nonastannide (4-) Anion Sn_9^{4-}, a novel capped antiprismatic configuration (C_{4v}). J Am Chem Soc 99:3313–3317
23. Wilson WL, Rudolph RW, Lohr LL et al (1986) Multinuclear NMR characterization of anionic clusters of the main-group elements Ge, Sn, Sb, Tl, Pb, and Bi in nonaqueous solution. Inorg Chem 25:1535–1541

24. Pons BS, Santure DJ, Taylor RC et al (1981) Electrochemical generation of the naked metal anionic clusters, $Sn_{9-x}Pb_x^{4-}$ (X = 0 to 9). Electrochim Acta 26:365–366
25. Teixidor F, Leutkens ML Jr, Rudolph RW (1983) Transition-metal insertion into naked cluster polyanions. J Am Chem Soc 105:149–150
26. Luetkens ML, Teixidor F, Rudolph RW (1984) Diheteronuclear and triheteronuclear metal cluster anions containing transition-metals. Inorg Chimi Acta 83:L13–L15
27. Rudolph RW, Wilson WL, Taylor RC (1981) Naked-metal clusters in solution. 4. Indications of the variety of cluster species obtainable by extraction of Zintl phases: Sn_4^{2-}, $TlSn_8^{5-}$, $Sn_{9-x}Ge_x^{4-}$ (x = 0–9) and $SnTe_4^{2-}$. J Am Chem Soc 103:2480–2481
28. Wilson WL (1982) Preperation and NMR characterization of tin and lead anionic clusters. PhD Dissertation, University of Michigan, Michigan
29. Spiekermann A, Hoffinann SD, Fässler TF (2006) The Zintl ion $[Pb_{10}]^{(2-)}$: a rare example of a homoatomic closo cluster. Angew Chem Int Ed 45:3459–3462
30. Bown M, Jelinek T, Stibr B et al (1988) Facile pathway-defined fluxional cluster isomerization in 10- Vertex Closo-2,1,6-metalladicarbaboranes of ruthenium and rhodium. J Chem Soc Chem Commun 974–975
31. Gupta U, Reber AC, Clayborne PA et al (2008) Effect of charge and composition on the structural fluxionality and stability of nine atom tin-bismuth Zintl analogues. Inorg Chem 47:10953–10958
32. Birchall T, Burns RC, Devereux LA et al (1985) A Sn-119 mossbauer study of some nonclassically (cluster) and classically bonded zintl anions. Inorg Chem 24:890–894
33. Rosdahl J, Fässler TF, Kloo L (2005) On the structure of nonastannide clusters in liquid and solid state. Eur J Inorg Chem:2888–2894
34. Campbell J, Mercier HPA, Holger F et al (2002) Snythesis, crystal structure and density functional theory calculations of the closo-$[1\text{-}M(CO)_3(E_9)]^{4-}$ cluster anions. Inorg Chem 41: 86–107
35. Kesanli B, Fettinger J, Eichhorn B (2001) The $[(\text{eta-}C_6H_5Me)NbSn_6Nb(\text{eta-}C_6H_5Me)]^{(2-)}$ ion: a complex containing a metal-stabilized $Sn_6^{(12-)}$ cyclohexane-like Zintl ion. Angew Chem Int Ed 40:2300
36. Rudolph RW, Wilson WL, Parker F et al (1978) Nature of naked-metal-cluster polyanions in solution. Evidence for $(Sn_{9-x}Pb_x)^{4-}$ (x = 0-9) and Sn-Sb Clusters. J Am Chem Soc 100: 4629–4630
37. Baudler M, Glinka K (1993) Monocyclic and polycyclic phosphanes. Chem Rev 93: 1623–1667
38. Rudolph RWT, Young RC (1979) The nature of naked metal clusters in solution. In: Tsutsui M (ed) Fundamental research in homogeneous catalysis, vol 3. Plenum, New York, pp 997–1005
39. Critchlow SC, Corbett JD (1981) Stable homopolyatomic anions – the Tetrastannide(2-) And Tetragermanide(2-) anions, $Sn_4^{(2-)}$ And $Ge_4^{(2-)}$ – X-ray crystal-structure of $K+(Crypt)_2Sn_4$. Ethylenediamine. J Chem Soc Chem Commun 5:236–237
40. Corbett JD, Edwards PA (1975) Stable homopolyatomic anions – crystal-structures of salts of anions Pentaplumbide(2-) and Enneastannide(4-). J Chem Soc Chem Commun 984–985
41. Wang LL, results to be published
42. Fässler TF, Hunziker M, Spahr ME et al (2000) Homoatomic clusters E_9^{3-} with E = Ge, Sn, and Pb: EPR spectra, magnetism and electrochemistry. Z Anorg Allg Chem 626:692–700
43. Kesanli B, Eichhorn BW, Fettinger JC (2001) The $[(\eta\text{-}C_6H_5Me)NbSn_6Nb(\eta\text{-}C_6H_5Me)]^{2-}$ complex containing an Sn_6^{12-} cyclohexane-like Zintl ion. Angew Chem Int Ed 40:2300–2302
44. Scharfe S, Fässler TF, Stegmaier S et al (2008) $[Cu@Sn_9]^{(3-)}$ and $[Cu@Pb_9]^{(3-)}$: Intermetalloid clusters with endohedral Cu atoms in spherical environments. Chem A Eur J 14:4479–4483
45. Wang JQ, Stegmaier S, Wahl B et al (2010) Step-by-step synthesis of the endohedral stannaspherene $Ir@Sn_{12}{}^{(3-)}$ via the capped cluster anion $Sn_9Ir(cod)^{(3-)}$. Chem A Eur J 16:1793–1798
46. Esenturk EN, Fettinger JC, Eichhorn BW (2006) Synthesis, structure, and dynamic properties of $[Ni_2Sn_{17}]^{(4-)}$. J Am Chem Soc 128:12–13

47. Kocak FS, Zavalij P, Lam YF et al (2008) Solution dynamics and gas-phase chemistry of $Pd_2@Sn_{18}^{(4-)}$. Inorg Chem 47:3515–3520
48. Kesanli B, Halsig JE, Zavalij P et al (2007) Cluster growth and fragmentation in the highly fluxional platinum derivatives of $Sn_9^{(4-)}$: synthesis, characterization, and solution dynamics of $Pt_2@Sn_{17}^{(4-)}$ and $Pt@Sn_9H^{3-}$. J Am Chem Soc 129:4567–4574
49. Kesanli B, Fettinger J, Gardner DR et al (2002) The $[Sn_9Pt_2(PPh_3)]^{2-}$ and $[Sn_9Ni_2(CO)]^{3-}$ complexes: two markedly different Sn_9M_2L transition metal zintl ion clusters and their dynamic behavior. J Am Chem Soc 124:4779–4786
50. Kocak FS, Moses M, Fettinger JC, Zavalij P et al (2010) $M@Sn_9H^{3-}$ and relation to fused deltahedron clusters (in prep)
51. Renner G, Kircher P, Huttner G et al (2001) Efficient syntheses of the complete set of compounds $[\{(OC)_5M\}_6E_6]^{(2-)}$ (M = Cr, Mo, W; E = Ge, Sn) – Structure and redox behaviour of the octahedral clusters $[Ge\text{-}_6]^{(2-)}$ and $[Sn_6]^{(2-)}$. Eur J Inorg Chem 973–980
52. Prabusankar G, Kempter A, Gemel C et al (2008) $Sn_{17}\{GaCl(ddp)\}_4$: A high-nuclearity metalloid tin cluster trapped by electrophilic gallium ligands. Angew Chem Int Ed 47:7234–7237
53. Rivard E, Steiner J, Fettinger JC et al (2007) Convergent syntheses of $Sn_7\{C_6H_3$–2, 6-$(C_6H_3$–2, 6-Pr-$i_2)_2{}^{-2}$: a cluster with a rare pentagonal bipyramidal motif. Chem Commun 46:4919–4921
54. Eichler BE, Power PP (2001) Synthesis and characterization of Sn8(2, 6-Mes(2)C6H3)4 (Mes=2, 4, 6-Me3C6H2): a main group metal cluster with a unique structure. Angew Chem Int Ed 40:796–797
55. Kesanli B, Fettinger J, Eichhorn B (2001) The closo-$[Sn_9M(CO)_3]^{(4-)}$ Zintl ion clusters where M = Cr, Mo, W: two structural isomers and their dynamic behavior. Chem A Euro J 7:5277–5285
56. Esenturk EN, Fettinger J, Eichhorn B (2006) The $Pb_{12}{}^{(2-)}$ and $Pb_{10}{}^{(2-)}$ zintl ions and the $M@Pb_{12}{}^{(2-)}$ and $M@Pb_{10}{}^{(2-)}$ cluster series where M = Ni, Pd, Pt. J Am Chem Soc 128:9178–9186
57. Esenturk EN, Fettinger J, and Eichhorn B (2005) The closo-$Pb_{10}{}^{(2-)}$ Zintl ion in the $[Ni@Pb_{10}]^{(2-)}$ cluster. Chem Commun 247–249
58. Eichhorn BW, Haushalter RC, Pennington WT (1988) Synthesis and structure of closo-$Sn_9Cr(CO)_3{}^{4-}$: the first member in a new class of polyhedral clusters. J Am Chem Soc 110:8704–8706
59. Eichhorn BW, Haushalter RC (1990) Closo-$[CrPb_9(CO)_3]^{4-}$: a 100 Year history of the non-aplumbide tetra-anion. J Chem Soc Chem Commun 937–938
60. Yong L, Hoffmann SD, Fässler TF (2005) Crystal structures of $[K(2.2.2\text{-crypt})]_4[Pb_9Mo(CO)_3]$ – isolation of the novel isomers $[(\eta^5\text{-}Pb_9)Mo(CO)_3]^{(4-)}$ beside $[(\eta^4\text{-}Pb_9)Mo(CO)_3]^{(4-)}$. Eur J Inorg Chem 3663–3669
61. Downing DO, Zavalij P, Eichhorn BW (2010) The closo- $Sn_9Ir(cod)^{(3-)}$ and $Pb_9Ir(cod)^{(3-)}$ Zintl ions: isostructural Ir-I derivatives of the nido-$E_9{}^{(4-)}$ anions (E = Sn, Pb). Eur J Inorg Chem 890–894
62. Charles S, Eichhorn BW, Bott SG et al (1996) Synthesis and characterization of $[P_7Ni(CO)]^{3-}$, $[HP_7Ni(CO)]^{2-}$ and $[E_7HPt(PPh_3)]^{2-}$. J Am Chem Soc 118:4713–4714
63. Sawyer AK, Brown YE, Hanson EL (1965) Di-n-butyltin halide hydrides. J Organomet Chem 3:464–468
64. Rivard E, Fischer RC, Wolf R et al (2007) Isomeric forms of heavier main group hydrides: experimental and theoretical studies of the Sn(Ar)H (2) (Ar=terphenyl) system. J Am Chem Soc 129:16197–16208
65. Esenturk EN, Fettinger J, Lam YF et al (2004) $[Pt@Pb_{12}]^{(2-)}$. Angew Chem Int Ed 43: 2132–2134
66. King RB, Heine T, Corminboeuf C et al (2004) Antiaromaticity in bare deltahedral silicon clusters satisfying Wade's and Hirsch's rules: an apparent correlation of antiaromaticity with high symmetry. J Am Chem Soc 126:430–431

67. Neukermans S, Janssens E, Chen ZF et al (2004) Extremely stable metal-encapsulated $AlPb_{10}^{+}$ and $AlPb_{12}^{+}$ clusters: mass-spectrometric discovery and density functional theory study. Phys Rev Let 92:163401
68. Schleyer PV, Maerker C, Dransfeld A et al (1996) Nucleus-independent chemical shifts: a simple and efficient aromaticity probe. J Am Chem Soc 118:6317–6318
69. Sun Z-M, Xiao H, Li J et al (2007) $Pd_2@Sn_{18}^{(4-)}$: Fusion of two enohedral stannaspherenes. J Am Chem Soc 129:9560–9561
70. Kocak FS, Zavalij PY, Lam YF et al (2009) Substituent-dependent exchange mechanisms in highly fluxional RSn_9^{3-} anions. Chem Commun 4197–4199
71. Burns RC, Corbett JD (1982) Heteroatomic polyanions of the post transition metals. The synthesis and structure of a copound containing Thallium nonastannide and Thallium Ocatastannide with a novel disorder. J Am Chem Soc 104:2804–2810
72. Critchlow SC, Corbett JD (1982) Heteropolyatomic anions of the post transition-metals – synthesis and structure of the Ditindibismuthide(2-) anion, $Sn_2Bi_2^{(2-)}$. Inorg Chem 21: 3286–3290
73. Hull MW, Sevov SC (2009) Functionalization of nine-atom deltahedral zintl ions with organic substituents: detailed studies of the reactions. J Am Chem Soc 131:9026–9037
74. Chapman DJ, Sevov SC (2008) Tin-based organo-Zintl ions: alkylation and alkenylation of $Sn_9^{(4-)}$. Inorg Chem 47:6009–6013
75. Hull MW, Sevov SC (2007) Addition of alkenes to deltahedral zintl clusters by reaction with alkynes: synthesis and structure of $[Fc\text{-}CH{=}CH\text{-}Ge_9CH\text{-}Fc]^{(2-)}$, an organo-zintl-organometallic anion. Angew Chem Int Ed 46:6695–6698
76. Hull MW, Ugrinov A, Petrov I et al (2007) Alkylation of deltahedral zintl clusters: synthesis of [R-Ge_9Ge_9-R](4-) (R = Bu-t, Bu-s, Bu-n, Am-t) and structure of [Bu-t-Ge_9-Ge_9-Bu-t](4-). Inorg Chem 46:2704–2708
77. Hull MW, Sevov SC (2007) Organo-zintl clusters soluble in conventional organic solvents: setting the stage for organo-zintl cluster chemistry. Inorg Chem 46:10953–10955
78. Ugrinov A, Sevov SC (2004) Rationally functionalized deltahedral zintl ions: synthesis and characterization of $[Ge_9\text{-}ER_3]^{(3-)}$, $[R_3E\text{-}Ge_9\text{-}ER_3]^{(2-)}$, and $[R_3E\text{-}Ge_9\text{-}Ge_9\text{-}ER_3]^{(4-)}$ (E = Ge, Sn; R = Me, Ph). Chem A Eur J 10:3727–3733
79. Schrenk C, Schellenberg I, Pottgen R et al (2010) The formation of a metalloid $Sn_{10}[Si(SiMe_3)_3]_6$ cluster compound and its relation to the alpha <->beta tin phase transition. Dalton Trans 39:1872–1876
80. Wang JQ, Stegmaier S, Fässler TF (2009) $[Co@Ge_{10}]^{3-}$: An intermetalloid cluster with archimedean pentagonal prismatic structure. Angew Chem Int Ed 48:1998–2002

Struct Bond (2011) 140: 91–131
DOI: 10.1007/430_2010_33

Published online: 5 January 2011

Relationships Between Soluble Zintl Anions, Ligand-Stabilized Cage Compounds, and Intermetalloid Clusters of Tetrel (Si–Pb) and Pentel (P–Bi) Elements

Thomas F. Fässler

Abstract The high abundance of nine- and seven- or eleven-atom clusters of group 14 and group 15 elements, respectively, led to an enhanced investigation of the versatile chemical reactivity of these species. The studies showed that these clusters are ideal precursors for the synthesis of larger units, functionalized cage molecules, and endohedrally filled clusters. They can further serve as building blocks for large homoatomic and heteroatomic intermetalloid clusters as well as for nanostructured materials. In this chapter, the chemistry and the structures of Zintl ions and their derivatives and of cage molecules are summarized and put in the context of intermetalloid clusters. The emphasis is put on the fact that irrespective of the starting material (Zintl ions or small low-valent organometallic compounds), the same types of intermetalloid clusters are formed, and the similarities of ligand-carrying *metalloid* clusters and *intermetalloid* anions that arise from reactions of *soluble* Zintl ions will be pointed out.

Keywords Cage compound · Cluster · Intermetalloid · Pentel · Structure · Tetrel · Zintl anion

Contents

Dedicated to Professor Dr. Hansgeorg Schnöckel on the Occasion of his 70th Birthday.

T.F. Fässler
Department of Chemistry, Technische Universität München, Lichtenbergstraße 4, 85747 Garching, Germany
e-mail: thomas.faessler@lrz.tum.de

1 Introduction and Remarks on Metalloid and Intermetalloid Clusters

Chemistry of metals and semimetals is diverse and still full of mysteries but at the same time holds a huge potential for applications. Metal- or semimetal-rich compounds also serve as excellent examples to follow the transition of matter from the bulk material to the molecule passing through the nanoscale regime. In a nano-sized metal cluster intermediate, electronic properties between the discrete state of a molecule and the transport properties of a bulk solid occur. Therefore, synthesis of such compounds is a high challenge. While the synthesis of nano-sized metal and semimetal clusters with a more and more narrow size distribution has been developed over the last decades, the controlled bottom-up synthesis of larger cluster units that are *well-defined* in composition and size is still unsatisfying.

In this chapter, we will describe bottom-up synthetic routes for cluster compounds which are rich in group 14 and 15 elements. Bare group 14 and 15 element clusters are well investigated in the gas phase, but scarcely exist in macroscopic amounts and are limited to simple molecules such as P_4, Sb_4, and Bi_4 and to carbon polymorphs such as C_{60} and C_{70}. When ligands attached to the atoms of the homoatomic cages are tolerated, a much larger variety of group 14 and 15 element clusters with homoatomic bonds is accessible, of which tetrahedrane $C_4{}^tBu_4$ [1], prismane, dodecahedrane, or functionalized fullerenes such as $C_{70}Cl_{16}$ [2] are pertinent examples in carbon chemistry, and $Si(Si\,{}^tBu_3)_4$ [3] and $Pb_{10}\{Si(SiMe_3)_3\}_6$ [4] are representatives of the heavier homologs (Fig. 1). Ligand-stabilized metal clusters are more common for group 13 elements which form larger cluster anions such as $[Al_{77}R_{20}]^{2-}$ [5] (Fig. 1) and $[Ga_{84}R_{20}]^{4-}$[6] [$R = N(SiMe_3)_2$]. These so-called metalloid clusters [7] have recently received a lot of attention since they represent the "missing link" between an extended solid (metal) and a soluble cluster.

An elegant access to larger cluster units is to make use of preformed smaller clusters as they occur, e.g., in Zintl phases [8, 9]. The phase KSi (or K_4Si_4) contains homoatomic Si_4 tetrahedra, and thus a salt-like formulation $(Na^+)_4[Si_4]^{4-}$ with a formal electron transfer is appropriate (Fig. 1) [10]. Such homoatomic anionic building blocks occur in binary or ternary intermetallic phases A_nE_m and A_nPn_m with A being an alkali and alkaline-earth metal, and E = Si–Pb and Pn = P–Bi. These phases are generally available in good quantities, and – since a few members

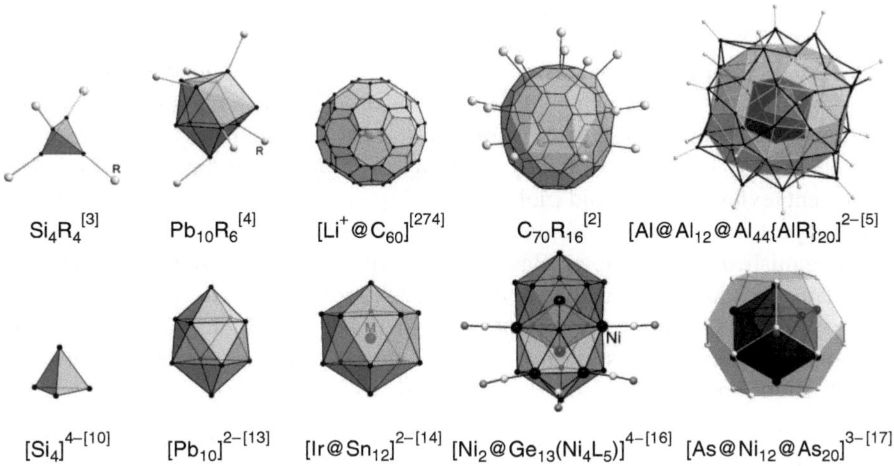

Fig. 1 Examples of polyhedral cage molecules and metalloid clusters (first row) compared to related Zintl ions and intermetalloid clusters that are obtained from the reaction of Zintl ions (second row)

are soluble in polar, organic solvents – they have a large potential as starting materials for the synthesis of larger clusters. The availability of such so-called Zintl ions in large quantity provides a much better starting position than all the improved synthetic approaches to fullerenes [11] or to group 13 element clusters [12]. The synthesis of homoatomic clusters with up to seven and nine atoms for group 15 and 14 elements, respectively, is relatively easy to accomplish and occurs by simply reacting the respective elements with alkali metals (A) in the melt. The solids that contain deltahedral tetrel (E) clusters $[E_9]^{4-}$ or pentel (Pn) cage anions $[Pn_7]^{3-}$ or $[Pn_{11}]^{3-}$ are soluble and therefore readily available for subsequent syntheses. Selected representatives that are obtained directly from Zintl phases or by subsequent reaction of soluble Zintl ions are shown in the second row of Fig. 1. By starting from soluble homoatomic group 14 and 15 polyanions cluster growth such as the formation of $[Pb_{10}]^{2-}$ [13] by fragmentation and oxidative coupling of Zintl ions, the introduction of organic ligands, the synthesis of endohedral cage anions such as $[Ir@Sn_{12}]^{2-}$ [14], and the formation of larger intermetalloid cluster anions [15] such as $[Ge_{13}Ni_7(CO)_5]^{4-}$ [16] or $[As_{21}Ni_{12}]^{3-}$ [17] can be achieved (Fig. 1). Many structures of the products thus obtained compare well to those of cage molecules or intermetalloid clusters which are made by step-by-step synthesis from small precursor molecules with only one or a few group 14 and 15 atoms. Several review articles have been published on group 14 [18–21] and group 15 clusters [22, 23], and a comprehensive list of compounds is available from a recent survey [24].

In this chapter, the chemistry and the structures of Zintl ions and cage molecules are summarized and put in the context of *intermetalloid* clusters. The emphasis is put on the observation that irrespective of the starting material (Zintl ions or small low-valent organometallic compounds) the same types of intermetalloid clusters are formed.

The terms *metalloid* and *intermetalloid* are used according to the following definitions: The term *metalloid* cluster was introduced for species that contain both ligand-bearing and ligand-free metal atoms which are bonded exclusively to other metal atoms, and which contain more metal–metal than metal–ligand bonds. With respect to their topology, metalloid clusters show similarities to the respective bare elements [7]. Zintl ions as they occur in neat Zintl phases have been excluded in a recent review on metalloid triel clusters [25], but *soluble* Zintl ions are included in this report since they are without metal–ligand bonds and bare many similarities as accomplished in this chapter. Based on the term *metalloid* cluster, the expression *intermetalloid* cluster was coined to describe Zintl ions which encapsulate a second type of metal [15]. For two reasons we will point out the similarity of ligand-carrying *metalloid* clusters to *intermetalloid* anions which arise from reactions of *soluble* Zintl ions:

(a) As will be shown in Chap. 3.2, the vinylation of the Zintl ion $[Ge_9]^{4-}$ leads to $[Ge_9R]^{3-}$ and $[Ge_9R_2]^{2-}$, whereas the anion $[Ge_9R_3]^{-}$ has been synthesized from low-valent molecular precursors and has been described as metalloid cluster. Although a direct synthesis of $[Ge_9R_3]^{-}$ by a threefold alkylation of a Zintl anion precursor has not been achieved yet, the three alkylated clusters bear the same metalloid characteristics.
(b) Reactions of the soluble Zintl ions $[Ge_9]^{4-}$, $[Sn_9]^{4-}$, and $[As_7]^{3-}$ with organometallic compounds lead to, e.g., $[Ni_2@Ge_{13}\{Ni_4(CO)_5\}]^{4-}$, $[Ir@Sn_{12}]^{2-}$, and $[As@Ni_{12}@As_{20}]^{3-}$, respectively (Fig. 1). All three examples have a central atom with coordination number 12 in a more or less distorted icosahedral environment, just as it is observed for Al in the metalloid anion $[Al@Al_{12}@Al_{44}\{AlN(SiMe_3)_2\}_{20}]^{2-}$. The structure of $[Ni_2@Ge_{13}\{Ni_4(CO)_5\}]^{4-}$ consists of two interpenetrating icosahedra (Ni_3Ge_9 and Ni_5Ge_7) as pointed out in Fig. 1 rather similar to the Pd_{33} core of three interpenetrating icosahedra constructed from Pd^0 atoms in $[Pd_{69}(CO)_{36}(PEt)_{14}]$ [26]. The structural motif of interpenetrating icosahedra, which is also found, e.g., in the α-Mn structure and its intermetallic derivatives [27], is a common motif of intermetallic compounds, and therefore the expression *intermetalloid* cluster is well grounded.

In the following chapters, all references to the compounds shown in the figures are given in the respective tables. General references and references to compounds not shown are given in the text.

2 Homoatomic Cluster and Cage Anions

2.1 Zintl Phases Containing Deltahedral Tetrel Clusters and Pentel Cage Anions

Zintl phases A_nE_m and A_nPn_m of the elements E = Si–Pb, Pn = P–Bi and A = Na–Cs can be described using the *Zintl–Klemm* formalism [28–30]. A formal electron transfer from the active metal A to the main group element E/Pn leads to a salt-like description with polyanions $[E/Pn_m]^{n-}$ and A^+ cations. In many cases, the structures of the polyanions can easily be derived by applying the valence concept and forming covalent bonds between the E/Pn atoms (8–N rule). In the case of deltahedral $[E/Pn_m]^{n-}$ clusters, electron-counting rules according to the Wade–Mingos rules are applicable [31].

Upon dissolution of Zintl phases in polar solvents such as liquid ammonia [32], ethylenediamine [33, 34], or dimethylformamide [35], a full charge separation can be achieved. Ion separation in soluble Zintl phases is additionally supported when sequestering agents for the cations such as [2.2.2]crypt = 4,7,13,16,21,24-hexaoxa-1,10-diazabicyclo-[8.8.8]hexacosane [36] or [18]crown-6 = 1, 4, 7, 10, 13, 16-hexaoxacyclooctadecane [37] and their derivatives are used. However, such a transition from a salt-like, semiconducting Zintl *phase* to an ionic Zintl *anion* only takes place for a small number of compounds, and the high abundance of nine- and seven- or eleven-atom clusters of group 14 and group 15 elements, respectively, led to an enhanced investigation of the versatile chemical reactivity of these species. The studies showed that they are ideal precursors for the synthesis of larger clusters, functionalized cage molecules, and endohedrally filled clusters, and they can serve as building blocks for large homoatomic and heteroatomic intermetalloid clusters or nanostructured materials.

2.2 Structures of Soluble Homoatomic Zintl Anions of Group 14 and 15 Elements

Smaller tetrel element clusters like the tetrahedral $[E_4]^{4-}$ ions were observed for the first time in the alloy NaPb [38] and are present in AE phases for E = Si to Pb and A = Na–Cs. Although these phases are not soluble, ammoniates of $[Pb_4]^{4-}$ have been obtained from liquid ammonia solutions of the binary phase RbPb [39]. The Zintl phases $A_{12}Si_{17}$ (A = K, Rb, Cs) and $K_6Rb_6Si_{17}$, which in the solid state contain $[Si_4]^{4-}$ and $[Si_9]^{4-}$ anions in the ratio 2:1, readily dissolve in liquid ammonia, and recently it has been shown that both cluster anions $[Si_4]^{4-}$ [37] and $[Si_9]^{4-}$ [40] are retained in solution. Corresponding pentel element clusters in bulk solids are $[Pn_7]^{3-}$ and $[Pn_{11}]^{3-}$ (Table 1).

Soluble Zintl ions can also be obtained by electrochemical methods using the respective element as cathode material [34, 60, 61], or through the reaction of the various modifications of the tetrel (Sn and Pb) and pentel (P, As, Sb) elements with dissolved or finely dispersed alkali or alkaline-earth metals in solution [62] as well as in molten crown-ethers [63].

Beside the most frequent soluble tetrel and pentel element clusters listed in Table 1, other homoatomic polyanions have been isolated from solution. An overview is given in Figs. 2 and 3 for group 14 and 15 elements, respectively, and in Table 2. The smaller tetrel clusters $[E_5]^{2-}$ and $[E_{10}]^{2-}$ have been isolated as A-crypt salts (Fig. 2b, e, respectively), and most probably they are formed upon

Table 1 Polyanions of tetrel and pentel elements of soluble Zintl phases

Polyanion	Composition	Ref.
$[Si_4]^{4-}/[Si_9]^{4-}$	$A_{12}Si_{17}$ (A = K, Cs)	[41]
	$Rb_{12}Si_{17}$	[42]
	$A_{12}Si_{17}$ (A = K, Cs)	[41]
$[Ge_9]^{4-}$	K_4Ge_9	[43]
	Cs_4Ge_9, Rb_4Ge_9	[44]
$[Ge_4]^{4-}/[Ge_9]^{4-}$	$A_{12}Ge_{17}$ (A = Na, Rb, Cs)	[41, 45]
$[Sn_5]^{2-}$	$K_{70}Sn_{103}$[a]	[46]
$[Sn_9]^{4-}$	K_4Sn_9	[47]
$[Sn_4]^{4-}/[Sn_9]^{4-}$	$A_{12}Sn_{17}$ (A = K, Rb, Cs)	[41, 45]
	$A_{52}Sn_{82}$ (A = K, Cs)	[48]
	$K_{70}Sn_{103}$[a]	[46]
	$Na_{12}Sn_{17}$	[41]
$[Pb_9]^{4-}$	A_4Pb_9 (A = K, Rb)	[41, 49]
	Cs_4Pb_9	[50]
$[P_7]^{3-}$	Li_3P_7	[51]
	A_3P_7 (A = Na, K, Rb)	[23]
	Cs_3P_7	[52]
$[P_{11}]^{3-}$	A_3P_{11} (A = Na, K, Rb, Cs)	[23]
$[As_7]^{3-}$	A_3As_7 (A = Li, Na)	[53]
	K_3As_7	[54]
	Rb_3As_7	[54, 55]
	Cs_3As_7	[54, 56]
$[As_{11}]^{3-}$	A_3As_{11} (A = Li, Na)	[57]
	A_3As_{11} (A = K, Rb, Cs)	[58]
$[Sb_7]^{3-}$	Cs_3Sb_7	[59]

[a] $K_{70}Sn_{103} = (K^+)_{70}([Sn_4]^{4-})_{11}([Sn_5]^{2-})([Sn_9]^{4-})_6$

fragmentation of the nine-atom clusters. As yet $[Pb_{10}]^{2-}$ is the only bare ten-atom cluster [13]. The analogous cluster $[Ge_{10}]^{2-}$ is described in a crystal structure with strong rotational disorder, but the same anion is ordered in the crystal structure of $[Ge_{10}\{Mn(CO)_5\}]^{3-}$ where it is bound to a transition metal carbonyl fragment [138]. Various oxidized monomeric clusters $[E_9]^{n-}$ ($n = 2, 3$) have been reported. They occur in the same reactions that lead to the formation of the fourfold negatively charged clusters, i.e., upon dissolution of A_4E_9 phases in an appropriate solvent, except for E = Si, where $A_{12}Si_{17}$ has been used as a starting material. In most cases, the oxidized species are obtained in the absence of additional oxidizing agents. Thus, solvent impurities and the formation of amides – for example, in ethylenediamine solutions – most likely are involved in the oxidation processes.

The bonding situation of all known polyhedral structures $[E_m]^{4-}$ ($m = 4$ and 9) and $[E_m]^{2-}$ ($m = 5$, 9, and 10) is best described as a delocalized electron-deficient system in analogy to the boranes. Wade's rules [31, 139] can be applied if the radial B–H bonds of a borane are formally substituted by a lone pair of electrons at each cluster vertex atom. Thus, each group 14 element contributes two of its four valence electrons to the cluster skeletal bonding. The tetrahedral clusters $[E_4]^{4-}$ (Fig. 2a) therefore possess $2n + 4 = 12$ skeletal electrons (ske). According to Wade's rules, these units build a *nido*-cluster derived from the trigonal pyramidal *closo*-cages

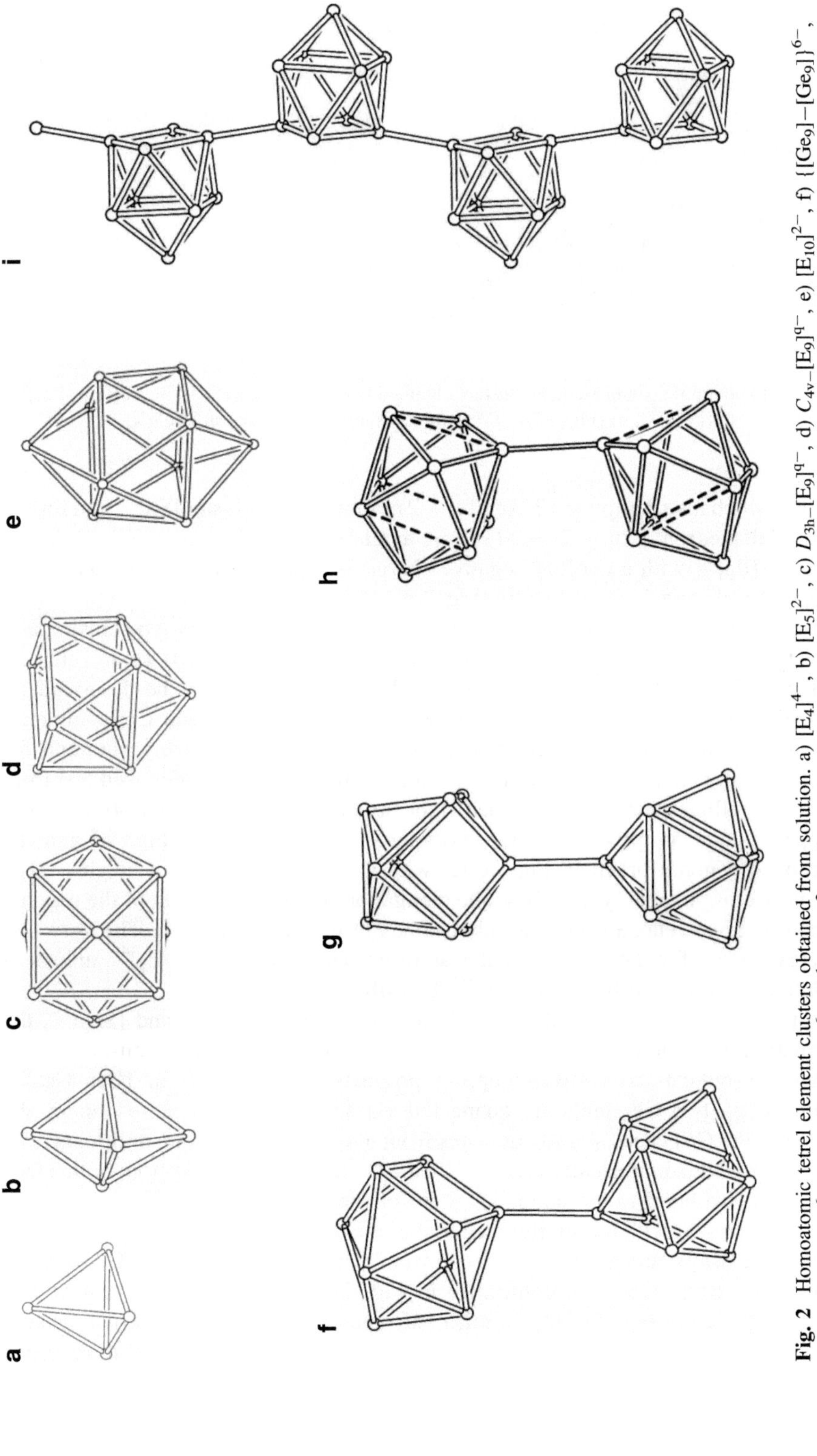

Fig. 2 Homoatomic tetrel element clusters obtained from solution. a) $[E_4]^{4-}$, b) $[E_5]^{2-}$, c) D_{3h}–$[E_9]^{q-}$, d) C_{4v}–$[E_9]^{q-}$, e) $[E_{10}]^{2-}$, f) $\{[Ge_9]-[Ge_9]\}^{6-}$, g) $\{[Ge_9]-[Ge_9]\}^{6-}$, h) $\{[Sn_9]-[Sn_9]\}^{6-}$, i) $^{1}_{\infty}\{[Ge_9]^{2-}\}$. Further examples are given in Table 2

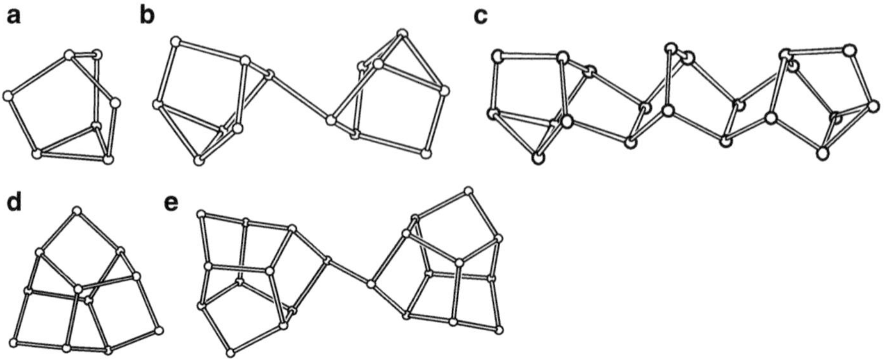

Fig. 3 Homoatomic pentel element clusters obtained from solution. a) $[Pn_7]^{3-}$, b) $[Pn_7{-}Pn_7]^{4-}$, c) $[Pn_{21}]^{3-}$, d) $[Pn_{11}]^{3-}$, e) $[Pn_{11}{-}Pn_{11}]^{4-}$. Further examples are given in Table 2

$[E_5]^{2-}$ which also comprise 12 ske ($2n + 2$). A nine-atom cluster $[E_9]^{2-}$ can form a *closo*-deltahedron if $2n + 2 = 20$ ske are available for the cluster bonding. Hence, clusters $[E_9]^{2-}$ with a twofold negative charge are appropriate to adopt the *closo*-shape of a tricapped trigonal prism (Fig. 2c) with point group symmetry D_{3h}. $[E_9]^{4-}$ clusters comprise $2n + 4 = 22$ ske which are required for a *nido*-type cluster, and thus form a C_{4v}-symmetric monocapped square antiprism (Fig. 2d). $[E_9]^{3-}$ clusters possess 21 electrons for their cluster bonding and thus cannot be described by Wade's rules, but their shape should integrate between D_{3h}- and C_{4v}-symmetric structures. Since distortions from idealized C_{4v}-symmetry are very common for $[E_9]^{4-}$ clusters in the solid state, their charges cannot reliably be deduced just from their crystallographically determined static structures [140]. The *nido*-type monocapped square antiprism with C_{4v} symmetry (II) as shown in Fig. 2d and the tricapped trigonal prism in Fig. 2c (I) with three elongated prism heights (D_{3h}) are very close in energy. Note that the elongation of only one height of the trigonal prism in I and relaxation of the other bonds lead to the cluster II. The very low energy barrier for the intramolecular atom exchange within the $[E_9]^{4-}$ anions in solution is evidenced by ^{119}Sn and ^{207}Pb NMR investigations.

In the case of the nine-atom tetrel element clusters $[Ge_9]^{4-}$ and $[Sn_9]^{4-}$, the formation of stable radicals $[E_9^{\bullet}]^{3-}$ compete with dimerization reactions, and aside from the radicals, oxidative coupling products are reported. In $[Ge_9{-}Ge_9]^{6-}$ (Fig. 2f, g), two Ge_9 units are connected via an *exo*-bond. In these dimers, the shape of the Ge_9 cages slightly deviates from a monocapped square antiprism. The open face of this unit usually is rhombohedrally distorted, and one atom of this face forms an *exo*-bond to an analogous atom of the second cluster. The intercluster bond is collinear with the shorter diagonal of the open face of each cluster (Fig. 2f), and the clusters are arranged in a *transoid* conformation. $K_6(Ge_9{-}Ge_9)(dmf)_{12}$ contains – aside from the conformer in Fig. 2f – another dimeric unit with a shape depicted in Fig. 2g [35]. In this conformer, both clusters are derived from a tricapped trigonal prism with two strongly elongated prism heights leading to two

Table 2 Structurally characterized homoatomic tetrel and pentel element cluster anions obtained from solutions

E	Si	Ge	Sn	Pb	Fig.
$[E_4]^{4-}$	$[Si_4]^{4-}$ [37][a]	–	$[Sn_4]^{4-}$ [39]	$[Pb_4]^{4-}$ [39]	2a
$[E_5]^{2-}$	$[Si_5]^{2-}$ [64]	$[Ge_5]^{2-}$ [65, 66]	$[Sn_5]^{2-}$ [67]	$[Pb_5]^{2-}$ [67, 68]	2b
$[E_9]^{q-}$	$[Si_9]^{2-}$ [69]	$[Ge_9]^{3-}$ [70–72]	$[Sn_9]^{3-}$ [71, 73–76]	$[Pb_9]^{3-}$ [71, 74, 77]	2c, d
($q = 2, 3, 4$)	$[Si_9]^{3-}$ [64]	$[Ge_9]^{4-}$ [78–82]	$[Sn_9]^{4-}$ [63, 83–90]		
	$[Si_9]^{4-}$ [40]			$[Pb_9]^{4-}$ [77, 86, 91, 92]	
$[(E_9)^{3-}]_2$	–	$[(Ge_9)–(Ge_9)]^{6-}$ [35, 93–95]	$[(Sn_9)–(Sn_9)]^{6-}$ [141]	–	2f, g
$[(E_9)^{2-}]_n$	–	$\{[Ge_9]^{2-}\}_3$ [96, 97]	–	–	6a
($n = 3, 4, \infty$)		$\{[Ge_9]^{2-}\}_4$ [98, 99]			6b
		${}^1_\infty\{[Ge_9]^{2-}\}$ [100, 101]			2i
$[E_{10}]^{2-}$	–	$[Ge_{10}]^{2-}$ [102]	–	$[Pb_{10}]^{2-}$ [13]	2e
Pn	**P**	**As**	**Sb**	**Bi**	
cyclo-$[Pn_n]^{q-}$	$[P_4]^{2-}$ [103–105]	$[As_4]^{2-}$ [105–107]	$[Sb_4]^{2-}$ [108]	$[Bi_4]^{2-}$ [109, 110]	
($n = 4, 5, 6, 8$)		$[As_6]^{4-}$ [111]	$[Sb_5]^{5-}$ [112]		
			$[Sb_8]^{8-}$ [113]		
$[Pn_7]^{3-}$	$[P_7]^{3-}$ [103, 114–119]	$[As_7]^{3-}$ [120–123]	$[Sb_7]^{3-}$ [108]	–	3a
$[Pn_7–Pn_7]^{4-}$	$[P_{14}]^{4-}$ [124, 125]	$[As_{14}]^{4-}$ [124]			3b
$[Pn_7–Pn_7–Pn_7]^{3-}$					
$[Pn_{11}]^{3-}$	$[P_{11}]^{3-}$ [116, 126–130]	$[As_{11}]^{3-}$ [131, 132]	$[Sb_{11}]^{3-}$ [132, 133]	–	3d
$[Pn_{11}–Pn_{11}]^{4-}$	$[P_{22}]^{4-}$ [134]	$[As_{22}]^{4-}$ [135]			
$[Pn_{21}]^{3-}$	$[P_{21}]^{3-}$ [136]	–	–	–	3c
$[Pn_{26}]^{4-}$	$[P_{26}]^{4-}$ [137]	–	–	–	

[a]Known only as MesCu adduct in $[(MesCu)_2Si_4]^{4-}$

vertex-sharing rectangular cluster faces. The *exo*-bonds between the two apex atoms approximately points to the cluster centers. A dimeric $[Sn_9–Sn_9]^{6-}$ anion has been obtained by oxidation of $[Sn_9]^{4-}$ with AgMes [141]. Its *cisoid* conformation (Fig. 2h) most probably arises from the coordination to the bridging Ag^+ ion (Fig. 7k). The $Ge_9–Ge_9$ and $Sn_9–Sn_9$ *exo*-bonds are slightly longer than the corresponding single bonds in the diamond modifications of the elements but are shorter than the distances to the nearest neighbors in the cluster.

Further oxidation of the nine-atom clusters to formal $[Ge_9]^{2-}$ leads to linear polymers $\{_\infty^1[\text{-}Ge_9\text{-}]^{2-}\}$ with two covalent intercluster bonds (Fig. 2i). Trimers $[Ge_9{=}Ge_9{=}Ge_9]^{6-}$(Fig. 6a) and tetramers $[Ge_9{=}Ge_9{=}Ge_9{=}Ge_9]^{8-}$ (Fig. 6b) occur via nonclassical bond formation between two neighboring atoms of the triangular prism basis planes of the *closo*-shaped clusters, which results in Ge–Ge–Ge bond angles of 90° and in considerably longer Ge–Ge contacts between the cluster units. Quantum-chemical calculations have shown that the *exo*-bonds participate in a delocalized electronic system that comprises the whole anion [204].

The polyanion $[Ge_{45}]^{12-}$ (Fig. 6c) represents a pentamer of $[Ge_9]^{4-}$ and arises from an oxidative coupling of five Ge_9 clusters in the reaction of an ethylenediamine solution of K_4Ge_9 and $Au(PPh_3)Cl$. The anion coordinates to four Au^+ and one K^+ cations. In the structure, four Ge_9 deltahedra are retained while the fifth has opened up to covalently link the four intact subunits.

Cationic deltahedral Pn clusters are known for Pn=Bi. $[Bi_5]^{3+}$, $[Bi_8]^{2+}$, and $[Bi_9]^{5+}$, which have been structurally characterized with complex counterions such as $AlCl_4^-$, AsF_6^-, and $HfCl_6^{2-}$, can be obtained either by molten salt routes or by using so-called super acidic systems [205–209]. According to Wade's formalism, $[Bi_5]^{3+}$ and $[Bi_8]^{2+}$ correspond to 12 ske *closo* and 22 ske *arachno* structures, respectively, and the latter adopts the shape of a square antiprism. $[Bi_9]^{5+}$ which is expected to form a 22 ske *nido*-cluster in analogy to Fig. 2d, rather adopts a distorted tricapped trigonal prismatic topology (Fig. 2c).

All other homoatomic Pn cluster anions are cage molecules with localized two-center-two-electron bonds. The cyclic $[Pn_4]^{2-}$ anion has been characterized by NMR spectroscopy for Pn = P [210], and structures are known for Pn = P, As, Sb, Bi (Table 2). A planar $[P_5]^-$ anion has only been detected in solution by ^{31}P NMR spectroscopy [210]. Envelop-shaped five- and six-membered rings are found for $[Pn_5]^{5-}$ and $[P_6]^{4-}$, whereas $[As_6]^{4-}$ deviates only slightly from planarity and has a chair conformation. The aromaticity of the planar anions $[P_4]^{2-}$ and $[P_6]^{4-}$ is distinct from that of regular 6π-aromatic hydrocarbons and has been described "lone-pair aromaticity" [103, 111]. An S_8-analogous, crown-shaped $[Pn_8]^{8-}$ unit has been obtained for Pn = Sb.

Soluble heptapnicanortricyclane anions $[Pn_7]^{3-}$ (Fig. 3a) and trishomocubane-shaped (ufosane-like) anions $[Pn_{11}]^{3-}$ (Fig. 3d) are very common and known as in the binary solids for Pn = P, As, Sb (Table 2). Oxidative coupling of these monomers leads to the dimers $[Pn_7–Pn_7]^{4-}$ and $[Pn_{11}–Pn_{11}]^{4-}$ for Pn = P and As (Fig. 3b, e), which – as observed for the tetrel element clusters – have an external homoatomic bond, but in this case the structures of the monomeric units are fully retained upon dimerization. A trimeric oxidative coupling product of $[P_7]^{3-}$ is the henicosaphosphide $[P_{21}]^{3-}$ (Fig. 3c) which contains a central norbornane-like P_7 unit that has been formed by opening of the three-membered ring in the monomer, which is connected to two heptaphosphanortricyclane P_7 cages via four covalent bonds. The largest Pn cluster known so far is a $[P_{26}]^{4-}$ unit found in $Li_4P_{26}(thf)_{16}$. It contains two heptaphosphanortricyclane P_7 cages that are covalently linked to a P_{12} unit which itself consists of two edge-fused norbornane-like P_7 units.

3 Zintl Anions, Ligand-Stabilized Cage Compounds, and Intermetalloid Clusters of the Tetrel Elements

3.1 Silicon

The as yet structurally characterized molecules and anions of silicon are listed in Table 3, and most of the representatives are shown in Fig. 4. The cage compounds Si_4R_4 (R = Si $^tBu_3$ and R = $SiMe\{CH(SiMe_3)_2\}_2$) and $Si_8(Si^tBu_3)_6$ both contain Si_4 tetrahedra, and the latter compound (Fig. 4d) attests that higher-coordinated and ligand-free Si atoms can coexist already in rather small molecules. The anion $[Si_4(SiMe\{CH(SiMe_3)_2\}_2)_3]^-$ also contains an Si_4 core with only three ligands. This remarkable anion shows ligand scrambling on the NMR time scale in solution and contains a bridging ligand in the solid state. The two examples shown in Fig. 4a, b can formally be described as the product of the addition of four and three alkyl cations R^+, respectively, to the Zintl ion $[Si_4]^{4-}$. The first proof of a soluble $[Si_4]^{4-}$ species has been provided by the isolation of $[Si_4(MesCu)_2]^{4-}$ (Fig. 4c) from a liquid ammonia solution of $[Si_4]^{4-}$, $[Si_9]^{4-}$, and MesCu. The distorted tetrahedral Si_4 anion is stabilized by two Cu atoms in bridging positions. $[(\eta^4\text{–}Si_9)Zn(C_6H_5)]^{3-}$ (Fig. 4e) and the Ni-bridged anion $[(Si_9)_2\{\mu^2\text{-}Ni(CO)_2\}_2]^{8-}$ (Fig. 4f) represent the only transition metal complexes of $[Si_9]^{4-}$ clusters. From an electronic point of view, the Zn species is a *closo* cluster in which a ZnR^+ fragment is attached to a $[Si_9]^{4-}$ unit.

Table 3 List of silicon cage compounds and clusters

Formula		Ref.	Fig.
Si_4R_4	R = Si $^tBu_3$	[3]	4a
	R = $SiMe\{CH(SiMe_3)_2\}_2$	[211]	
$[Si_4R_3]^-$	R = $SiMe\{CH(SiMe_3)_2\}_2$	[211]	4b
$[Si_4(ML)_2]^{4-}$	ML = CuMes	[37]	4c
Si_8R_6	R = Si^tBu_3	[212]	4d
$[Si_9\text{–}MR]^{3-}$	MR = $Zn(C_6H_5)$	[154]	4e
$[Si_9\text{–}\{ML_2\}_2\text{–}Si_9]^{8-}$	ML_2 = $Ni(CO)_2$	[213]	4f

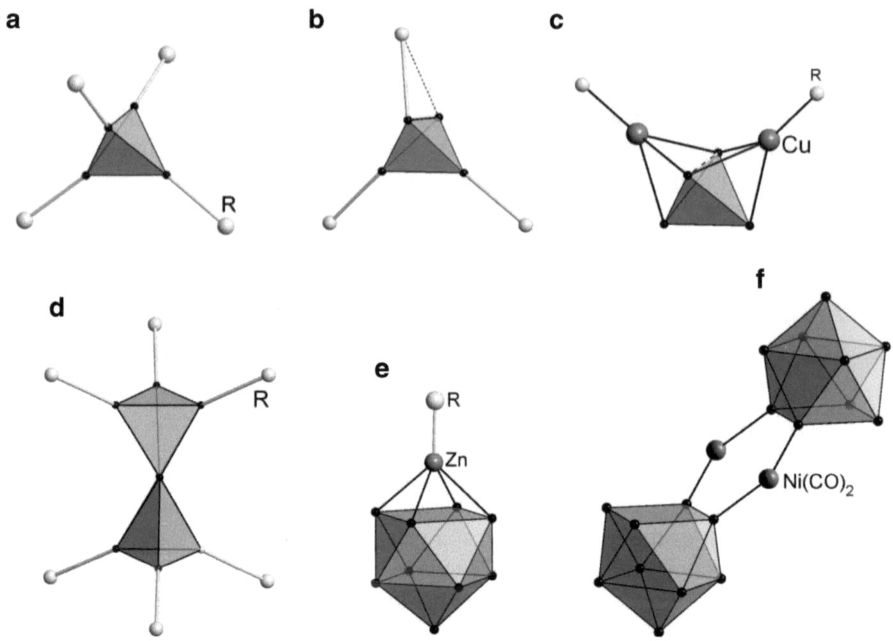

Fig. 4 Silicon cage compounds and clusters. a) Si_4R_4, b) $[Si_4R_3]^-$, c) $[Si_4\ (ML)_2]^{4-}$, d) Si_8R_6, e) $[Si_9\text{-}MR]^{3-}$, f) $[Si_9\text{-}\{ML_2\}_2\text{-}Si_9]^{8-}$. Further examples are given in Table 3

3.2 Germanium

The large number of cage compounds and deltahedral clusters of germanium are listed in Table 4, and most examples are shown in Figs. 5 and 6. With an increasing number of Ge atoms, the cage molecules contain less ligands than cage atoms. In the tetrahedral molecules Ge_4R_4 all Ge atoms carry a ligand R (Fig. 5a), whereas the molecules Ge_5R_4, Ge_6R_2 (Fig. 5b), and Ge_8R_6 (Fig. 5c) contain one, four, and two ligand-free Ge atoms, respectively. An octahedral $[Ge_6]^{2-}$ subunit has been found in the transition metal complexes $[\{GeM(CO)_5\}_6]^{2-}$. In their structure, which is analogous to the one shown in Fig. 7b, each Ge cluster atom is coordinated to an $M(CO)_5$ fragment (M = Cr, Mo, and W). Regarding solely the cluster core $[Ge_6]^{2-}$ unbosoms the electronic relationship to Ge_6R_6 by merely addition of two R^+ ligands.

In the same way leads formal addition of R^+ units to the Zintl ion $[Ge_9]^{4-}$ to the series $[Ge_9R_n]^{(4-n)-}$. The derivatives are known for $n = 1$, 2 (Fig. 5d, e) and 3 (Table 4). Monosubstituted $[Ge_9\text{-}Mes]^{3-}$ has been synthesized just recently using Mes-Ag as an alkylation reagent, and a functionalization with ferrocenyl ligands has also been realized for R = C(H)=C(H)(Cp_2Fe). Functionalized dimers $[(R\text{-}Ge_9)\text{–}(Ge_9\text{-}R)]^{6-}$ occur as well, even though the mechanism of their formation is not fully understood. With respect to the bisfunctionalized monomers $[R\text{–}Ge_9\text{–}R]^{2-}$, the dimers can be regarded as their heterosubstituted derivatives

Table 4 List of germanium cage compounds and clusters

Formula		Ref.	Fig.
Ge_4R_4	$R = Si\,^tBu_3$	[171]	5a
Ge_5R_4	$R = CH(SiMe_3)_2$	[172]	
Ge_5R_4	$R = 2,6\text{-}(2,4,6\text{-}Me_3C_6H_2)_2\text{-}C_6H_3$	[172]	
Ge_6R_2	$R = 2,6\text{-}(2,6\text{-}^iPr_2C_6H_3)_2\text{-}C_6H_3$	[142]	5b
$[Ge_6(ML_5)_6]^{2-}$	$M = Cr; L = CO$	[143]	
	$M = Mo, W; L = CO$	[144]	
Ge_8R_6	$R = N(SiMe_3)_2$	[173]	5c
	$R = C_6(O^tBu)_2H_3$	[174]	
$[Ge_9R]^{3-}$	$R = CH{=}CH_2$	[175]	5d
	$R = SnMe_3, SnPh_3$	[176]	
	$R = C(CH_3){=}CH\text{–}CH_2\text{–}CH_3$	[177]	
$[Ge_9R_2]^{2-}$	$R = CH{=}CH_2$	[178]	5e
	$R = CD{=}CD_2, CH_2\text{–}CH(CH_2)_2$	[177]	
	$R = C(CH_3){=}CH\text{–}CH_2\text{–}CH_3$	[177]	
	$R = CH{=}CHFc$	[179]	
	$R = SbPh_2, BiPh_2$	[180]	
	$R = GePh_3, SnMe_3, SnPh_3$	[176]	
$[Ge_9RR']^{2-}$	$R = C_6H_5, R' = SbPh_2$	[181]	5e
	$R = SbPh_2, R' = Ge_9(SbPh_2)^{2-}$		
	$R = {}^tBu, R' = Ge_9(^tBu)^{2-}$	[182]	
	$R = SnPh_3, R' = Ge_9(SnPh_3)^{2-}$	[176]	
$[Ge_9R_3]^-$	$R = Si(SiMe_3)_3$	[183]	5f
$[Ge_9R_3(CrL_5)]^-$	$R = Si(SiMe_3)_3, L = CO$	[184]	5g
$[Ge_9R_3(CrL_3)]^-$	$R = Si(SiMe_3)_3, L = CO$	[184]	5h
$[Ge_9\text{–}ML]^{3-}$	$M = Cu, L = P^iPr_3$ and PCy_3	[185]	5i
	$M = Ni, L = CO$	[186]	
	$M = Pd, L = PPh_3$	[187]	
$[Ge_9\text{–}MR]^{3-}$	$M = Zn, R = C_6H_5$	[154]	5i
	$R = {}^iPr$, Mes	[155]	5i
$[Ge_9\text{–}M(Ge_9)]^{7-}$	$M = Cu$	[185]	5j
$[Ge_9M_3Ge_9]^{5-}$	$M = Au$	[188]	5k
$[M(Ge_9R_3)_2]^-$	$M = Cu, Ag; R = Si(SiMe_3)_3$	[189]	5l
	$M = Au; R = Si(SiMe_3)_3$	[190]	
$[M(Ge_9R_3)_2]^-$	$M = Zn, Cd, Hg; R = Si(SiMe_3)_3$	[191]	5l
$^1_\infty([MGe_9]^{2-})$	$M = Hg$	[192, 193]	5m
$[M_3(Ge_9)_4]^{10-}$	$M = Hg$	[194]	
$[Ge_{10}\{ML_4\}]^{3-}$	$M = Mn, L = CO$	[138]	5n
$[Ge_{10}\{ML_4\}_8]^{6-}$	$M = Fe, L = CO$	[195]	5o
$[Ge_{10}R_6I]^+$	$R = Si^tBu_3$	[196]	5p
$[Ge_{10}(SiR_3)_4(SiR_2)R']^-$	$R = SiMe_3, R' = Me$	[197]	
$[Ge_{14}R_5]^{3-}$	$R = Ge(SiMe_3)_3$	[198]	5q
$[M_3Ge_{45}]^{9-}$	$M = Au$	[199]	6c
$[M@Ge_9]^{3-}$	$M = Ni$	[200]	6d
$[M@Ge_9\text{–}M'L]^{2-}$	$M = M' = Ni, L = CO$	[186]	6e
	$M = M' = Ni, L = PPh_3$	[16]	
	$M = Ni, M' = Pd, L = PPh_3$	[187]	
$[M@Ge_9\text{–}M'L]^{3-}$	$M = M' = Ni, L = CCPh$	[186]	
$[M_3@Ge_{18}]^{4-}$	$M = Ni$	[200]	6f
$[M_2@Ge_{18}]^{4-}$	$M = Pd$	[201]	6g
$[M_2@Ge_{13}M_4L_5]^{4-}$	$M = Ni, L = CO$	[16]	6h
$[M@Ge_{10}]^{3-}$	$M = Co$	[202]	6i
	$M = Fe$	[203]	

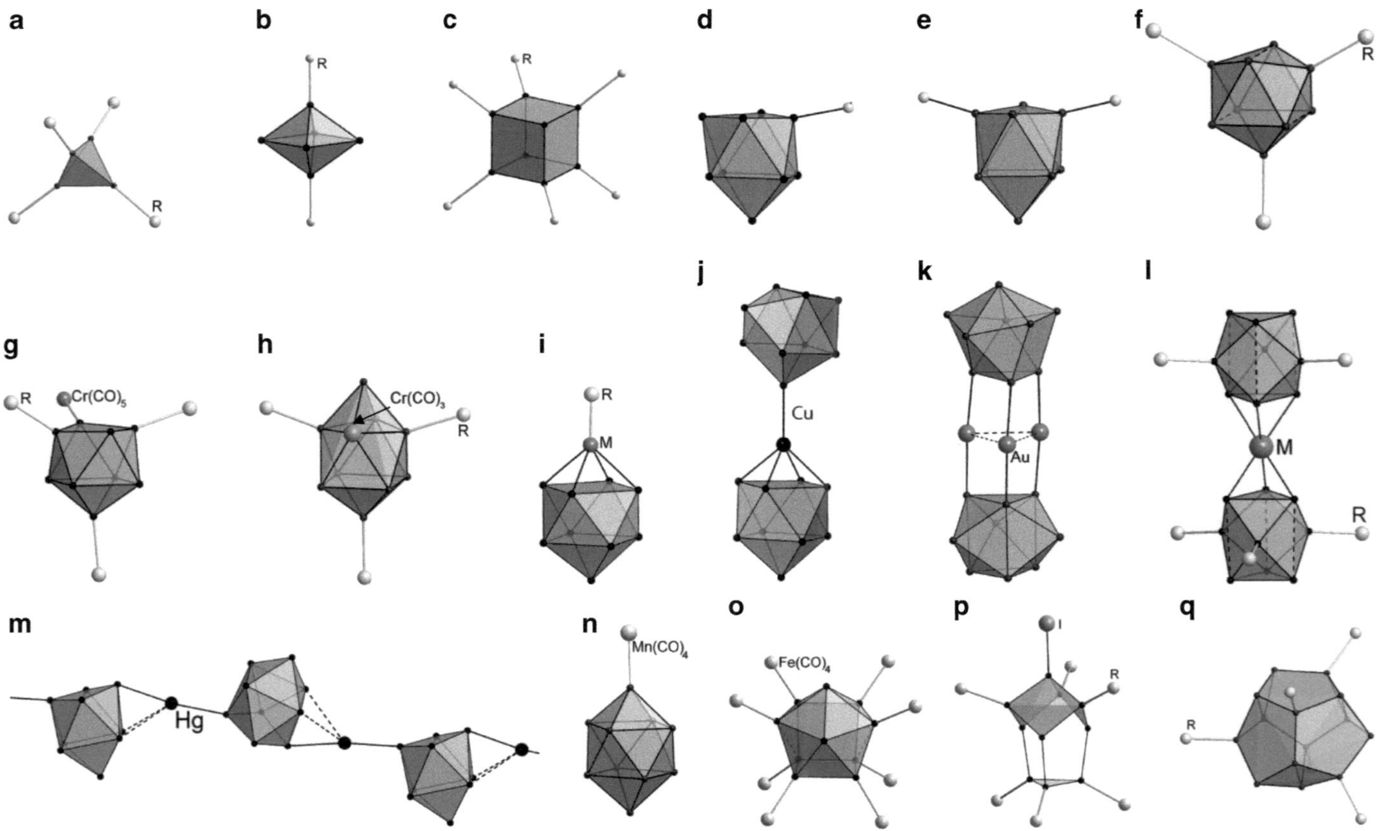

Fig. 5 Cage compounds and clusters of germanium. a) Ge_4R_4, b) Ge_6R_2, c) Ge_8R_6, d) $[Ge_9R]^{3-}$, e) $[Ge_9R_2]^{2-}$, f) $[Ge_9R_3]^{-}$, g) $[Ge_9R_3(CrL_5)]^{-}$, h) $[Ge_9R_3(CrL_3)]^{-}$, i) $[Ge_9ML]^{3-}$, j) $[(\eta^4\text{-}Ge_9)Cu(\eta^1\text{-}Ge_9)]^{7-}$, k) $[Ge_9Au_3Ge_9]^{5-}$, l) $[M(\eta^3\text{-}Ge_9R_3)_2]^{-}$, m) $^1_\infty\{[HgGe_9]^{2-}\}$, n) $[Ge_{10}\{ML_4\}]^{3-}$, o) $[Ge_{10}\{ML_4\}_8]^{6-}$, p) $[Ge_{10}R_6I]^{+}$, q) $[Ge_{14}R_5]^{3-}$. Further examples are given in Table 4

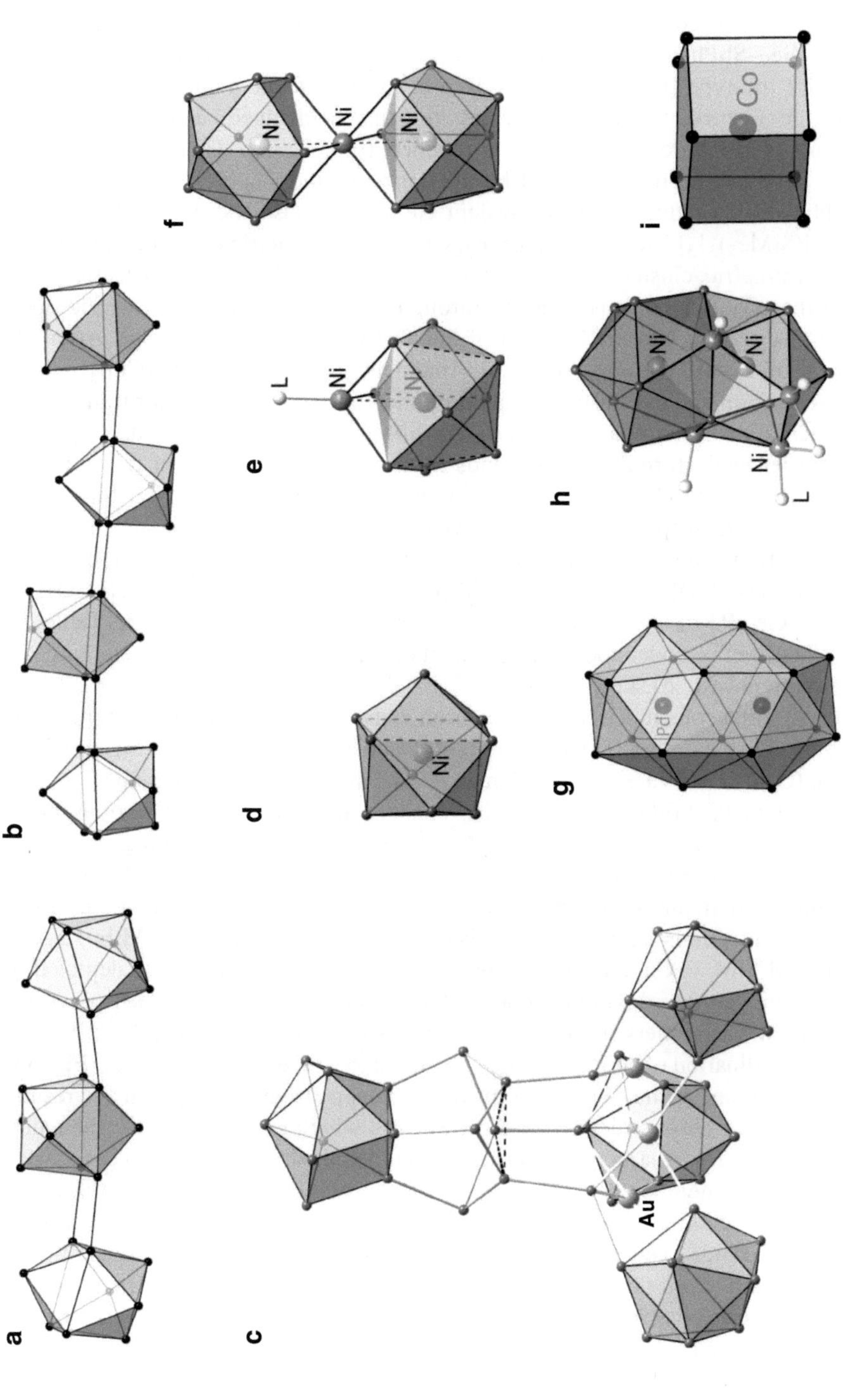

Fig. 6 Linked, endohedral, and intermetalloid germanium clusters. a) $[Ge_9{=}Ge_9{=}Ge_9]^{6-}$, b) $[Ge_9{=}Ge_9{=}Ge_9{=}Ge_9]^{8-}$, c) $[Au_3Ge_{45}]^{9-}$, d) $[Ni@Ge_9]^{3-}$, e) $[M@Ge_9\_M'L]^{2-}$, f) $[Ni_3@Ge_{18}]^{4-}$, g) $[Pd_2@Ge_{18}]^{4-}$, h) $[M_2@\ Ge_{13}M_4L_5]^{4-}$, i) $[M@Ge_{10}]^{3-}$. Further examples are given in Table 2 and 4

$[(\text{R–Ge}_9)\text{–R}']^{2-}$ with $\text{R}' = (\text{Ge}_9\text{R})^{2-}$. Two *exo*-bonded main group fragments can be introduced by oxidation of $[\text{Ge}_9]^{4-}$ with SbPh_3, which leads to $[\text{Ph}_2\text{Sb–Ge}_9\text{–SbPh}_2]^{2-}$ and $[\text{Ph}_2\text{Sb–(Ge}_9\text{–Ge}_9)\text{–SbPh}_2]^{4-}$. Further examples of dimers of the type $[\text{R–(Ge}_9\text{–Ge}_9)\text{–R}]^{4-}$ are given in Table 4. If exclusively *exo*-bonds are formed, the number of ske remains unchanged, and the *nido*-type structure of the bare Zintl cluster is retained (Fig. 5d, e). Trisalkylated mono-anions have not been synthesized by alkylation of the Ge_9 unit as yet, but they are obtained via reduction of low-valent Ge compounds. The deltahedral cluster $[\text{Ge}_9\{\text{Si(SiMe}_5)_3\}_3]^-$ with 22 ske appears as a D_{3h}-symmetric cage (Fig. 5f), while a C_{4v}-symmetric cluster is observed in $[(\eta^1\text{–Ge}_9)\{\text{Si(SiMe}_3)_3\}_3\{\text{Cr(CO)}_5\}]^-$ in which the $[\text{Ge}_9\text{R}_3]^-$ unit is attached through a ligand-free Ge atom of the open square of the cluster to the Cr atom of the Cr(CO)_5 fragment (Fig. 5g). After the removal of two CO groups from this fragment, the remaining Cr(CO)_3 unit is capable to form a vertex of the deltahedron thereby joining the cluster framework, and a ten-atom *closo* cluster $[(\eta^5\text{-Ge}_9\{\text{Si(SiMe}_3)_3\}_3)\text{Cr(CO)}_3]^-$ results (Fig. 5h). This cluster again formally corresponds to the threefold addition of R^+ to the η^5 isomers $[(\eta^5\text{–E}_9)\text{M(CO)}_3]^{4-}$ which are, however, known only for E = Sn and Pb (Figs. 7j and 9e, respectively). There exist a number of complexes which are η^4-capped by transition metal fragments at the open square of the Ge_9 cluster, of which $[(\eta^4\text{–Ge}_9)\text{Zn(C}_6\text{H}_5)]^{3-}$, $[(\eta^4\text{–Ge}_9)\text{Cu(PR}_3)]^{3-}$ (R = ^{i}Pr, Cy), $[(\eta^4\text{–Ge}_9)\text{Ni(CO)}]^{3-}$, and $[(\eta^4\text{–Ge}_9)\text{Pd(PPh}_3)]^{3-}$ have been structurally characterized (Fig. 5i; Table 4). All of them can be understood in the context of the isolobal concept as heteroatomic *closo* clusters with 20 ske.

The simultaneous function of a transition metal atom as a lone-pair acceptor and as a cluster vertex atom is realized in $[(\eta^1\text{–Ge}_9)\text{Cu}(\eta^4\text{–Ge}_9)]^{7-}$ (Fig. 5j), in which the two $[\text{Ge}_9]^{4-}$ units are connected in an η^1- and an η^4-fashion to the same Cu^+ ion. Transition metal bridges between Ge_9 clusters are also found in various other coordination modes: in $[\text{Ge}_9(\mu^2\text{–Au})_3\text{Ge}_9]^{5-}$ (Fig. 5k), three bridging Au atoms each are coordinated by two Ge atoms of two different Ge_9 units in an almost linear way and entertain aurophilic interactions within the resulting triangle of Au^+ ions; two $[\text{Ge}_9\text{R}_3]^-$ anions are coordinated via their triangular face to M^{2+} (M = Zn, Cd, Hg) and M^+ (M = Cu, Ag, Au) atoms, respectively, in the complexes depicted in Fig. 5l, and the coordination mode of the mercury atoms in Hg-Ge_9 polymers (Fig. 5m) and oligomers $[\text{Hg}_3(\text{Ge}_9)_4]^{10-}$ ranges from η^1 to η^3 (Table 4).

The coordination of an $[\text{Mn(CO)}_4]^-$ fragment to a bare $[\text{Ge}_{10}]^{2-}$ cluster (Fig. 5n) takes place at one of the capping atoms of the bicapped square antiprism and results in the expected *closo* structure for the Ge_{10} unit which is also found for the anion $[\text{Pb}_{10}]^{2-}$ (Fig. 2e). A structural transition from deltahedral structures to cages with four- and five-membered homoatomic rings is observed for functionalized Ge clusters. The anion $[\text{Ge}_{10}(\text{Fe(CO)}_4)_8]^{6-}$ (Fig. 5o) can be described as a $[\text{Ge}_{10}]^{6-}$ anion which binds to eight 16-electron Fe(CO)_4 fragments. If the two longer bonds (drawn as dashed lines in Fig. 5o) are considered as part of the cluster framework, a convex polyhedron with only two open squares is obtained, which resembles a strongly distorted *arachno*-type structure, derived from an icosahedron by the removal of two vertices and formation of rectangular faces. The $[\text{Ge}_{10}]^{6-}$ unit in

$[Ge_{10}(Fe(CO)_4)_8]^{6-}$ possesses 26 ske as required for a ten-atom *arachno* cluster, and the five-membered rings of an icosahedron are retained around those two Ge atoms which are not connected to Fe atoms. This clusters can also be described as a hybrid intermediate between a deltahedral cluster and a cube ("centaur polyhedron," i.e., half icosahedron and half cube, such as depicted in Fig. 5c). The cation $[Ge_{10}R_6I]^+$ (Fig. 5p) and the structurally related anion $[Ge_{10}(SiR_3)_4(SiR_2)R']^-$ also represent such a transformation from a deltahedral cluster to a not strictly deltahedral cage molecule. They possess one deltahedral face but also concave faces with both rectangles and five-membered rings. Most interestingly, the same connectivity of Ge atoms found in $[Ge_{10}R_6I]^+$ is partly also observed in the Ge_{45} unit shown in Fig. 6c.

As a general rule the number of four- and five-membered rings increases with an increasing number of Ge atoms in the cage, as documented, e.g., for $[Ge_{14}\{Ge(SiMe_3)_3\}_5]^{5-}$ in Fig. 5q.

As already mentioned above, the oxidation of the Ge_9 clusters leads to paramagnetic $[Ge_9^{\bullet}]^{3-}$ units or to the dimer $[Ge_9–Ge_9]^{6-}$ (Fig. 2f, g), and upon further oxidation the linear polymer $\{{}^1_\infty[-Ge_9-]^{2-}\}$ is formed (Table 2). In the dimeric cluster, the monomeric units are linked by a classical two-center-two-electron bond, whereas in the trimers and tetramers (Fig. 6a, b) these units are interconnected by nonclassical bonds. Until now cluster oligomerization products up to five Ge_9 clusters have been obtained. The multifaceted bonding in the corresponding $[Ge_{45}]^{12-}$ unit displayed in Fig. 6c includes covalent two-center-two-electron *exo*-cluster bonds as well as delocalized multicenter bonds in the deltahedral subunits. Long Ge–Ge contacts within a triangle of pentavalent Ge atoms correspond to a three-center-two-electron bond. As depicted in Fig. 6c, the Ge_{45} unit is coordinated to three Au^+ ions, each of which is interacting in a distorted square-planar fashion with four Ge atoms, thus leading to a final composition of $[Au_3Ge_{45}]^{9-}$.

In all examples described so far, the transition metal is coordinated in the outer sphere of the polyhedron. However, there also exists an impressive number of endohedrally filled cages (Table 4). The smallest representative of a transition metal-filled Zintl cluster is $[Ni@Ge_9]^{3-}$ (Fig. 6d). Although the structure refinement of this unit suffers from intense disorder, it can be concluded that the cluster adopts the shape of a strongly distorted tricapped trigonal prism with unequally elongated prism heights. According to EPR measurements, the endohedral nickel complex is paramagnetic and can be described as $[Ni^0@(Ge_9)^{3-}]$. Such endohedrally filled clusters can also act as ligands for transition metal fragments, as found, e.g., in $[(Ni@Ge_9)(NiL)]^{3-}$ (L = C≡C–Ph), $[(Ni@Ge_9)(NiL)]^{2-}$ (L = CO and PPh_3) (Fig. 6e; Table 4), and $[(Ni@Ge_9)Pd(PPh_3)]^{2-}$.

The intermetalloid cluster $[Ni_3@Ge_{18}]^{4-}$ (Fig. 6f) can be described as containing a Ni^0 atom that carries two $[Ni@Ge_9]^{2-}$ units (the reduced form is shown in Fig. 6d). The resulting cluster contains a linear chain of three Ni atoms and two widely opened Ge_9 polyhedra. These two Ge_9 units have a relative orientation that allows – after subsequent intercluster Ge–Ge bond formation – the formation of the ellipsoidally shaped cluster $[Pd_2@Ge_{18}]^{4-}$ shown in Fig. 6g. These anions are obtained by reacting $[Ge_9]^{4-}$-containing solutions, and thus an oxidative merging of two $[Ge_9]^{4-}$ clusters must occur during their formation. In the $[Pd_2@Ge_{18}]^{4-}$

anion, the two Pd^0 atoms are encapsulated by 18 tetrel atoms, which form a prolate deltahedral cluster, and the interstitial Pd atoms with a d^{10} electron configuration approximately occupy the ellipse focuses. Even though the Pd–Pd distance is rather short (2.831 Å), no Pd–Pd contacts are discussed. Another example of an endohedrally filled, ellipsoidally shaped cage is found in $[(Ni_2)@Ge_{13}\{Ni_4(CO)_5\}]^{4-}$ (Fig. 6h). This fully deltahedral cluster is built by 13 Ge and 4 Ni atoms, and two Ni atoms are encapsulated. Overall five CO ligands complete the coordination sphere of the outer Ni atoms. The cluster consists of two interpenetrating icosahedra with a common pentagonal face, so that each cap of this pentagon becomes the center of the neighboring icosahedron. As pointed out in the introduction, such interpenetrating icosahedra are indeed structural motifs of intermetallic phases and also of intermetalloid transition metal clusters such as $[Pd_{69}(CO)_{36}(PEt_3)_{18}]$, which consist of a Pd_{33} core of three interpenetrating Pd_{12} icosahedra [26].

With the anion $[Co@Ge_{10}]^{3-}$, the family of endohedral atoms was recently extended also to electron-poor d-block elements (Fig. 6i). As in the empty but ligand-stabilized clusters displayed in Fig. 5n–p, the structure of $[Co@Ge_{10}]^{3-}$ contains four- and five-membered rings in form of a pentagonal prism. The Ge–Ge and Ge–Co distances are comparable to that of binary intermetallic compounds, and the results of quantum-chemical calculations reveal a certain degree of localized Ge–Ge bonds. However, the electron count does not follow any conventional rules, and the coordination sphere of the transition metal atom in this cluster is more typical of an intermetallic phase [202]. The isosteric cluster $[Fe@Ge_{10}]^{3-}$ was discovered shortly later. It is expected to be paramagnetic, but this has not been verified yet.

3.3 Tin

The mixed-atom molecules $Sn_4Ge_2R_2$ contain a six-atom cage which forms a compressed octahedron (Fig. 7a), and the octahedral polyanion $[Sn_6]^{2-}$ (Fig. 7b) has been found as a ligand to ML_5 transition metal fragments (Table 5). Seven- and eight-atom polyhedra have been detected in gas-phase experiments but are less abundant and have not been observed yet as Zintl ion clusters. However, they exist as non-deltahedral cages and adopt the shape of a pentagonal bipyramide in Sn_7R_2 (Fig. 7c), and strongly and less distorted cubes have been found in Sn_8R_4 and Sn_8R_6, respectively (Fig. 7d, e). A cyclohexane-like Zintl ion is stabilized by η^3-coordination of two $Nb(C_5H_5Me)$ units in $[(C_5H_5Me)Nb(Sn_6)Nb(C_5H_5Me)]^{2-}$ [145]. The structure (not shown) contains a $[Sn_6]^{12-}$ Zintl ion which requires fully oxidized Nb^{5+} ions for charge compensation. The Nb_2Sn_6 core forms a strongly distorted cube. Thus the $[Sn_6]^{12-}$ ring is analogous to that of the crown-shaped (S_8-like) anion $[As_8]^{8-}$ found in the complexes $[As_8Nb]^{3-}$ [221] and $[As_8Mo]^{2-}$ (Fig. 11d) [230].

Alkylation and vinylation or the formation of other derivatives of the $[Sn_9]^{4-}$ polyanions exclusively lead to monosubstituted cluster anions (Fig. 7f). ^1H, ^{13}C, or ^{119}Sn NMR experiments on ^{i}Pr- and $SnCy_3$-substituted Sn_9-derivatives reveal a

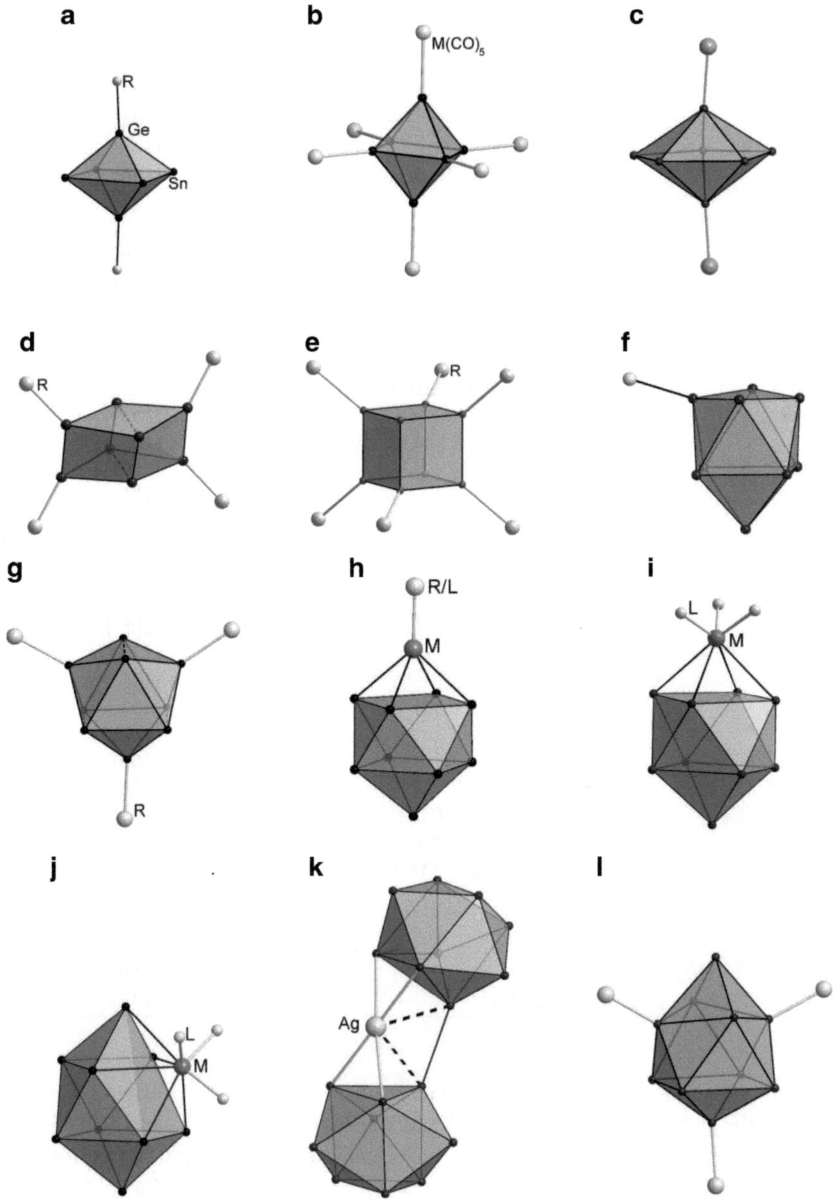

Fig. 7 Tin cage compounds and cluster complexes. a) $Ge_2Sn_4R_2$, b) $[Sn_6(ML_5)_6]^{2-}$, c) Sn_7R_2, d) Sn_8R_4, e) $[Sn_8R_6]^{2-}$, f) $[Sn_9R]^{3-}$, g) $[Sn_9R_3]$, h) $[(\eta^4\text{-}Sn_9)M\text{-}R)]^{3-}$, i) $[(\eta^4\text{-}Sn_9)ML_3]^{4-}$, j) $[(\eta^5\text{-}Sn_9)ML_3]^{4-}$, k) $[Ag(Sn_9\text{-}Sn_9)]^{5-}$, l) $[Sn_{10}R_3]^{+}$. Further examples are given in Table 5

dynamic behavior of these functionalized clusters in solution. A rearrangement of the Sn atoms within the Sn_9 framework leads to an exchange of the ligand position for R = ^{i}Pr, and a ligand migration is observed for R = $SnCy_3$. Functionalized

Table 5 List of tin cage compounds and clusters

Formula		Ref.	Fig.
$Ge_2Sn_4R_2$	$R = 2,6\text{-}(2,6\text{-}{}^iPr_2\text{–}C_6H_3)_2\text{–}C_6H_3$	[142]	7a
$[Sn_6(ML_5)_6]^{2-}$	M = Cr; L = CO	[143, 144]	7b
	M = Mo, W; L = CO	[144]	
$[Sn_6(ML)_2]^{2-}$	M = Nb, L = C_5H_5Me	[145]	
Sn_7R_2	$R = 2,6\text{-}(2,6\text{-}{}^iPr_2C_6H_3)_2C_6H_3$	[146]	7c
	R = GaCl(ddp)	[147]	7c
Sn_8R_4	$R = C_6H_3\text{-}2,6\text{-}(2,4,6\text{-}Me_3C_6H_2)_2$	[148]	7d
$[Sn_8R_6]^{2-}$	$R = Si^tBu_3$	[149]	7e
Sn_8R_6	$R = Si^tBu_3$	[149]	
$[Sn_9R]^{3-}$	$R = {}^iPr$	[150]	7f
	$R = {}^tBu$, CH=CH$_2$, CH=CHPh	[151]	
	$R = SnCy_3$	[150]	
$[Sn_8GeR]^{3-}$	R = CH=CH$_2$, CH=CH^{C}Pr	[152]	
$[Sn_7Ge_2R_2]^{2-}$	R = CH=CH$_2$, CH=CHPh	[152]	
Sn_9R_3	$R = 2,6\text{-}(2,4,6\text{-}{}^iPr_3C_6H_2)_2C_6H_3$	[153]	7g
$[(\eta^4\text{–}Sn_9)M\text{–}R]^{3-}$	M = Zn, R = C_6H_5	[154]	7h
	$R = C_3H_7, C_9H_{11}$	[155]	
	M = Cd, R = Ph, Sn(^{n}Bu)$_3$	[156]	
$[(\eta^4\text{–}Sn_9)ML]^{3-}$	M = Ir, L = cod	[14, 157]	7h
$[(\eta^4\text{–}Sn_9)ML_3]^{4-}$	M = Cr, L = CO	[158, 159]	7i
	M = Mo, L = CO	[159–161]	
	M = W, L = CO	[208]	
$[(\eta^5\text{–}Sn_9)ML_3]^{4-}$	M = W, L = CO	[159]	7j
$[M(Sn_9)_2]^{5-}$	M = Ag	[141]	7k
$[Sn_{10}R_3]^+$	$R = 2,6\text{-}(2,4,6\text{-}Me_3C_6H_2)_2C_6H_3$	[153]	7l
$Sn_{10}R_6$	$R = Si(SiMe_3)_3$	[162]	
$D_{3h}\text{–}[M@Sn_9]^{3-}$	M = Cu	[163]	8a
$C_{4v}\text{–}[M@Sn_9]^{3-}$	M = Cu	[164]	8b
$[(M@Sn_9)M'L]^{3-}$	M = M′ = Ni, L = CO	[165]	8c
	M = M′ = Pt, L = PPh$_3$	[165]	8d
$[Sn@\{Sn_8(SnR)_6\}]$	$R = N(2,6\text{-}{}^iPr_2C_6H_3)(SiMe_2R')$, R′ = Me, Ph	[166]	8e
$[M@Sn_{12}]^{3-}$	M = Ir	[14]	8f
$[M_2@Sn_{17}]^{4-}$	M = Ni	[167]	8g
	M = Pt	[168]	8h
$[Pd_2@Sn_{18}]^{4-}$		[169, 170]	8i
$[Sn_{17}R_4]$	R = GaCl(ddp)	[147]	8j

nine-atom clusters with a mixed Sn–Ge framework are also known. The vinyl groups in the mono- and di-functionalized clusters $[Sn_8GeR]^{3-}$ and $[Sn_7Ge_2R]^{2-}$ in both cases are attached to the Ge atom. All molecules and polyanions with deltahedral tin cages Sn_n observed for $n = 6$, 7, 9, and 10 (Fig. 7a–c, f–l) follow Wade's rules. Interestingly, the trialkylated and paramagnetic $[Sn_9R_3]^0$ molecule corresponds to a 21 ske cluster and thus is electronically equivalent to the paramagnetic $[Sn_9]^{3-}$ polyanions (Fig. 2c). The $[Sn_{10}R_3]^+$ unit (Fig. 7l) formally represents a trisalkylated *closo*-$[Sn_{10}]^{2-}$ cluster, but it has not been synthesized via the addition of three R^+-residues to an $[Sn_{10}]^{2-}$ unit, and the bare Sn_{10} cluster unit has also not been obtained as yet. Disproportionation of SnBr in solution and in the presence of $LiSiR_3$ (R = $SiMe_3$) leads to the cage molecule $Sn_{10}R_6$ (not shown) with a structure resulting from the interpenetration of a cube and an icosahedron. The shape of the Sn_{10} polyhedron is in perfect analogy to the one of the Ge_{10} cage in the anion $[Ge_{10}(Fe(CO)_4)_8]^{6-}$ shown in Fig. 5o. Assuming that an SnR unit and a ligand-free Sn vertex contribute three and two electrons for skeletal bonding, respectively, $Sn_{10}R_6$ has 26 ske which – as pointed out above – also holds for $[Ge_{10}(Fe(CO)_4)_8]^{6-}$.

The capping of the $[Sn_9]^{4-}$ cluster by a transition metal fragment $M(CO)_3$ was the first reaction in which a Zintl ion became attached to a transition metal. It was carried out in 1988 by *Haushalter* for M = Cr [158]. The reaction has subsequently been extended to M = Mo and W, and to the isolobal capping fragments Ir(cod), and M–R with M = Zn and Cd (Table 5). The ML_3, ML, and MR fragments generally coordinate to the open square of the nine-atom cluster and thus form *closo* clusters. For $ML_3 = W(CO)_3$, an isomerization of the *closo* cluster also leads to an η^5-Sn_9 complex shown in Fig. 7j. The *closo*-type structure of $[(\eta^5\text{-}Sn_9)M(CO)_3]^{4-}$ has the same number of ske as the trisfunctionalized $[(\eta^5\text{-}Ge_9)\{Si(SiMe_3)_3\}_3Cr(CO)_3]^-$ unit shown in Fig. 5h. Oxidative coupling of Sn_9 clusters can be achieved with the aid of MesAg as oxidizing agent. The newly formed Sn–Sn bond that connects the two monomeric units is supported by a silver atom according to the structure shown in Fig. 7k.

There exist several examples with late transition metals as endohedral atoms in Sn cages. The anion $[Cu@Sn_9]^{3-}$ can adopt two different structures with C_{4v}- and D_{3h}-symmetry, which are shown in Fig. 8a, b. In the latter, the nine tin atoms surround the enclosed Cu atom in an almost spherical manner. The cluster anion shows dynamic behavior in solution, as proven by ^{119}Sn and ^{63}Cu NMR spectroscopy. The compound is diamagnetic, and thus the Cu atom has the oxidation state +1. Both cluster types although with different encapsulated metals are also found as a ligand in transition metal complexes, as shown in Fig. 8c, d. An η^4 coordination mode occurs in the Ni(CO) complex of $[Ni@Sn_9]^{3-}$, and the observed η^3 coordination of the $Pt(PPh_3)$ fragment to $[Pt@Sn_9]^{3-}$ enforces an opening of the trigonal face of the basic trigonal prism in the resulting anion $\{[\eta^3\text{-}(Pt@Sn_9)]Pt(PPh_3)\}^{3-}$ (Fig. 8d). An endohedral tin atom is found inside a Sn_{14} cage which is built up from eight Sn atoms without heteroatomic contacts to external ligands and six Sn atoms bound to alkyl groups (Fig. 8e). The central tin atom has a remarkable coordination number of 14. While this Sn_{14} cluster has been obtained by the reduction of low-valent tin alkyls, the endohedral 12-atom tin cluster $[Ir@Sn_{12}]^{3-}$ is formed through the oxidation of $[(\eta^4\text{–}Sn)Ir(cod)]^{3-}$ (Fig. 8f) in a solution containing also dppe. According to DFT calculations, the Ir atom is negatively polarized, and the formal charge allocation $[Ir^{1-}@(Sn_{12})^{2-}]$ accounts for a *closo* Sn_{12} cluster endohedrally filled with a d^{10} transition metal.

Other clusters with more than nine tin atoms are shown in Fig. 8g–i. They have been obtained from ethylenediamine solutions of K_4Sn_9 and $Ni(cod)_2$, $Pt(PPh_3)_4$, and $Pd(PPh_3)_4$, respectively. $[Ni_2@Sn_{17}]^{4-}$ (Fig. 8g) formally consists of two $[Ni@Sn_9]^{2-}$ units in the shape of a tricapped trigonal prism with two elongated prism heights, which share the Sn atom that caps the two elongated prism heights. The two cluster units are rotated 90° with respect to each other, and the Sn_{17}

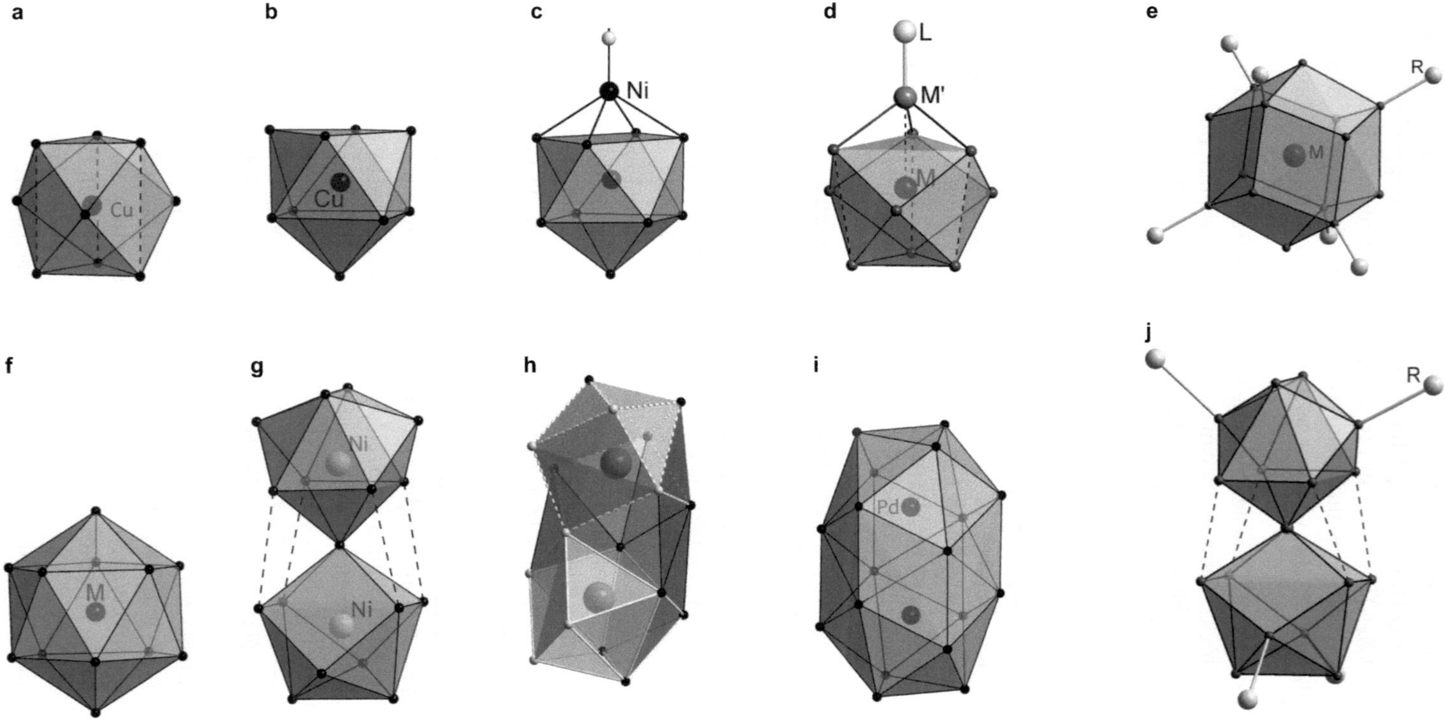

Fig. 8 Intermetalloid clusters of tin. a) D_{3h}–$[Cu@Sn_9]^{3-}$, b) C_{4v}–$[Cu@Sn_9]^{3-}$, c) $[(Ni@Sn_9)Ni(CO)]^{3-}$, d) $[(Pt@Sn_9)Pt(PPh_3)]^{3-}$, e) $[Sn@\{Sn_8(SnR)_6\}]$, f) $[Ir@Sn_{12}]^{3-}$, g) $[Ni_2@Sn_{17}]^{4-}$, h) $[Pt_2@Sn_{17}]^{4-}$, i) $[Pd_2@Sn_{18}]^{4-}$, j) $[Sn_{17}R_4]$. Further examples are given in Table 5

polyhedron has point group symmetry D_{2d}. The connecting Sn atom is surrounded by eight other Sn atoms. In solution, the Sn atoms of this cluster are fluxional on the NMR time scale. A second isomer of this 17-atom polyhedron with a rather different shape has been found in the isovalence electronic cluster anion $[Pt_2@Sn_{17}]^{4-}$ (Fig. 8h). In this structure, two cluster units that derive from a C_{4v}-symmetric *nido*-type nine-atom cluster (shown with light-gray solid and dashed lines, respectively) share two atoms, and the two rectangular open faces of the C_{4v}-symmetric building block appear at the cluster surface. An ellipsoidal deltahedron with a rather similar shape is found in $[Pd_2@Sn_{18}]^{4-}$ (Fig. 8i), which is isostructural to the above-mentioned $[Pd_2@Ge_{18}]^{4-}$ unit (Fig. 6g). Again the two Pd^0 atoms are encapsulated by 18 E atoms, but this time with an even longer Pd–Pd distance.

The ligand-stabilized intermetalloid cluster $[Sn_{17}\{GaCl(ddp)\}_4]$ has almost the same shape as the ligand-free cluster unit in $[Ni_2@Sn_{17}]^{4-}$. The Sn_{17} skeleton shown in Fig. 8j contains four Sn atoms with *exo*-bonds to the ligands. Structurally these four atoms correspond to the Sn atoms that cap the distorted trigonal prisms in the two $[Ni@Sn_9]^{2-}$ units in Fig. 8g. Although the prisms in Fig. 8g are empty, both clusters have rather similar Sn–Sn distances. Assuming that the two central Ni atoms are Ni^0 and thus do not contribute electrons to skeletal bonding in Fig. 8g, and that the cluster in Fig. 8j can be regarded as the product of a formal addition of four positively charged ligands to the anion, the same cluster framework for $[Ni_2@Sn_{17}]^{4-}$ and $[\square@Sn_{17}]^{4-}$ results, indicating that the number of ske seems to be decisive for the shape of the intermetalloid clusters.

3.4 Lead

In analogy to the corresponding tin clusters, the lead derivatives $[(\eta^4\text{–}Pb_9)M(CO)_3]^{4-}$ (M = Cr, Mo, W; Fig. 9d), $[(\eta^5\text{–}Pb_9)M(CO)_3]^{4-}$ (M = Mo; Fig. 9e), $[(\eta^4\text{–}Pb_9)MR]^{3-}$ (M = Zn and Cd), and $[(\eta^4\text{–}Pb_9)Ir(cod)]^{3-}$ (Fig. 9c) are known. For the endohedral $[Cu@Pb_9]^{3-}$ cluster (Fig. 9b), only one isomer is reported, and $[M@Pb_{12}]^{2-}$ (M = Pd and Pt; Fig. 9h) appears to be isostructural to the stannaspherane in Fig. 8f (Table 6).

A unique pentanuclear lead complex $[Pb_5\{Mo(CO)_3\}_2]^{4-}$ (Fig. 9a) is formed from a Zintl anion in solution. It contains a planar Pb_5 ring that coordinates two $\{Mo(CO)_3\}$ fragments in an η^5 fashion. Thus, according to quantum-chemical calculations the anion consists of an aromatic 2π electron system $[Pb_5]^{2-}$ and two $[Mo(CO)_3]^-$ units, and the Pb_5 anion is as an analog of the *cyclo*-Pn_5 unit that acts as a ligand in transition metal complexes shown in Fig. 11b and described below. The planar 2π electron five-membered ring is an interesting alternative to the deltahedral *closo* $[Pb_5]^{2-}$ cluster: While the electron deficiency in the deltahedral Pb_5 unit follows the delocalized picture according to Wade's rules, delocalized π-bonding is involved in the planar Pb_5 unit. Direct cluster coupling as observed

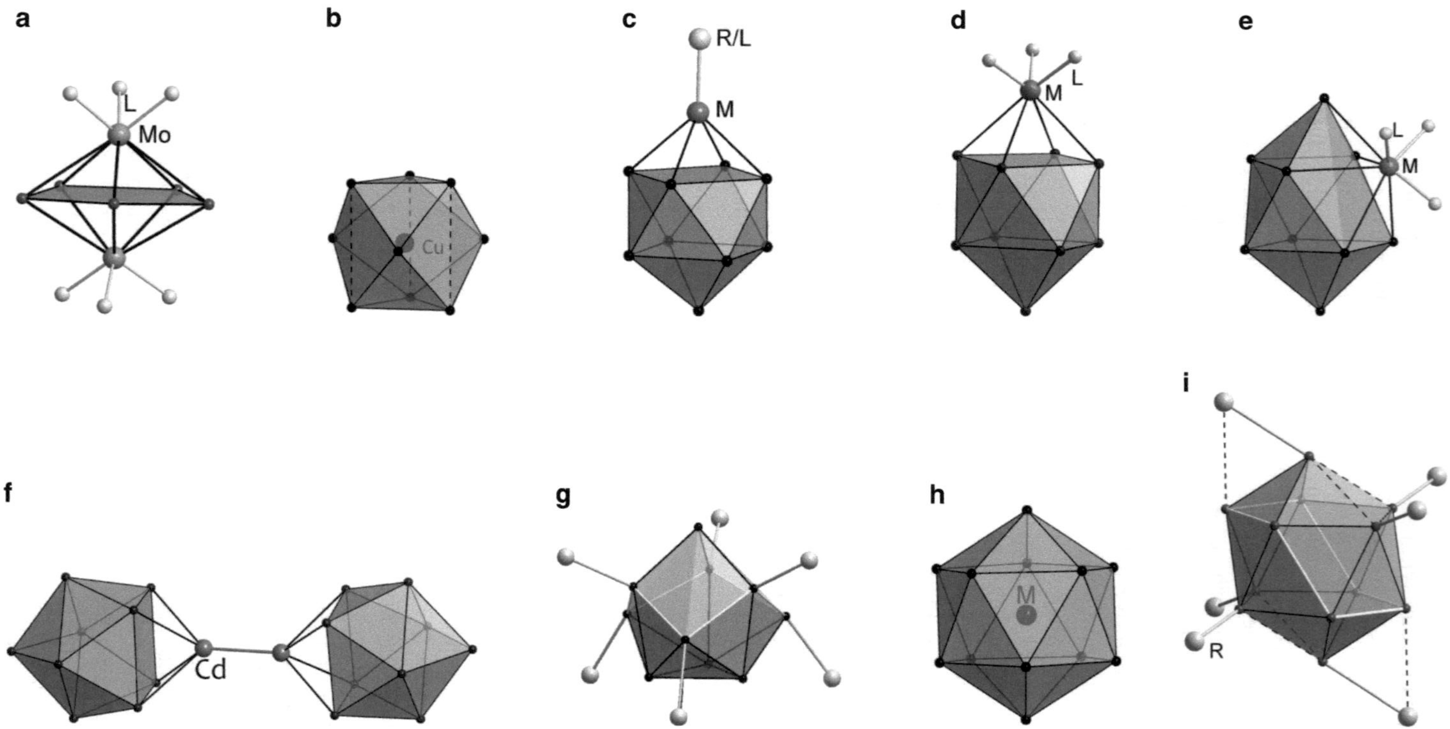

Fig. 9 Cage compounds, clusters complexes and intermetalloid clusters of lead. a) $[Pb_5\{Mo(CO)_3\}_2]^{4-}$, b) D_{3h}–$[Cu@Pb_9]^{3-}$, c) $[(\eta^4\text{–}Pb_9)MR]^{3-}$, d) $[(\eta^4\text{–}Pb_9)ML_3]^{4-}$, e) $[(\eta^5\text{–}Pb_9)Mo(CO)_3]^{4-}$, f) $[Pb_9\text{–}Cd\text{–}Cd\text{–}Pb_9]^{6-}$, g) $[Pb_{10}R_6]$, h) $[M@Pb_{12}]^{2-}$, i) $[Pb_{12}R_6]$. Further examples are given in Table 6

Table 6 List of structurally characterized tin cage compounds and clusters

Formula		Ref.	Fig.
$[Pb_5\{ML_3\}_2]^{4-}$	M = Mo, L = CO	[214]	9a
D_{3h}–$[M@Pb_9]^{3-}$	M = Cu	[163]	9b
$[(\eta^4\text{–}Pb_9)MR]^{3-}$	M = Zn, R = C_6H_5	[154]	9c
	M = Zn; R = C_7H_3, C_9H_{11}	[155]	
	M = Cd, R = C_6H_5	[156]	
$[(\eta^4\text{–}Pb_9)ML]^{3-}$	M = Ir, L = cod	[157]	9c
$[(\eta^4\text{–}Pb_9)ML_3]^{4-}$	M = Cr, L = CO	[215]	9d
	M = Mo, L = CO	[160, 216]	
	M = W, L = CO	[160]	
$[(\eta^5\text{–}Pb_9)ML_3]^{4-}$	M = Mo, L = CO	[216]	9e
$[Pb_9\text{–}Cd\text{–}Cd\text{–}Pb_9]^{6-}$		[217]	9f
$[Pb_{10}R_6]$	R = $Si(SiMe_3)_3$	[4]	9g
$[M@Pb_{10}]^{2-}$	M = Ni	[218, 219]	
$[M@Pb_{12}]^{2-}$	M = Ni, Pd, Pt	[219, 220]	9h
$[Pb_{12}R_6]$	R = $Si(SiMe_3)_3$	[4]	9i

for Ge_9 and Sn_9 clusters has not been found for Pb_9 units as yet, but two Pb_9 clusters are connected via two capping Cd atoms in $[(Pb_9)Cd\text{–}Cd(Pb_9)]^{6-}$ shown in Fig. 9f. The two highly charged Pb_9 ligands are obviously able to stabilize an otherwise rather unstable covalent Cd–Cd bond.

Ligand-stabilized lead cages are, e.g., $Pb_{10}(Si\{SiMe_3\}_3)_6$ and $Pb_{12}(Si\{SiMe_3\}_3)_6$ shown in Fig. 9g and i, respectively. In the former, the structure of the Pb_{10} unit deviates strongly from a deltahedron and contains four vertex-sharing rectangular faces. The cluster can also be described as a $(PbR)_6$ hexagon (white bonds in Fig. 9g) with a chair conformation to which one Pb atom that binds to three ring atoms is attached from above and a Pb_3 triangle from below. Alternatively the cage can be regarded as a 26 ske *hypo*-type cluster $[Pb_9R_6]^{2-}$ which is coordinated to a Pb^{2+} cation. The cage topology is related to one of the Ge_{10} unit in the anion $[Ge_{10}(Fe(CO)_4)_8]^{6-}$ shown in Fig. 5o and to that of the Sn_{10} cage in the isovalence electronic compounds $Sn_{10}R_6$ which both have a structure resulting from the interpenetration of a cube and an icosahedron. The structure further shows a great similarity to that of the cluster unit in the endohedral anion $[Ni@(Ge_9NiL)]^{2-}$ of Fig. 6e. The cubic part of the cages of the two lighter homologs displays two distorted squares and two squares with one shorter diagonal (rhombs), whereas in the Pb_{10} polyhedron three distorted squares and three squares with shorter diagonals (rhombs) are present.

A neutral cage with a distorted I_h-symmetric Pb_{12} cage and six *exo*-bonded ligands is found in $Pb_{12}(Si\{SiMe_3\}_3)_6$ (Fig. 9i). The six lead atoms *without* substituent form a ring with a chair conformation (solid gray bonds) as inversely observed for the Pb–R units in Fig. 9g. Each of the remaining six Pb atoms binds to a hypersilyl group with slightly longer bonds between these Pb atoms (dashed lines). Even though the Pb_{12} unit resembles an icosahedron, the cluster does not follow Wade's rules for deltahedra as it is the case for the undistorted ligand-free icosahedra with an endohedral Pd or Pt atom.

4 Zintl Anions, Ligand-Stabilized Cage Compounds, and Intermetalloid Clusters of the Pentel Elements

4.1 Phosphorous

The chemistry of homoatomic group 15 element anions is dominated by the formation of localized bonds between the Pn atoms rather than by cluster formation with delocalized electronic states. This arises from the larger number of valence electrons available. Since twofold connected Pn atoms in a group 15 element polyanion can reach their octet of electrons by only one additional electron – whereas two electrons are required for achieving the same bonding situation in the case of group 14 element polyanions – the resulting cluster anions carry relatively low negative charges. However, this situation changes drastically with an increasing number of metal atoms attached to a Pn polyanion, as will be shown below.

The only deltahedral phosphorus cage known so far is P_4. This unit may in exceptional cases stay intact upon coordination [232] to a transition metal, as observed in $[Cu(\eta^2\text{–}P_4)_2]^+$ [233], but in principle its reactivity is dominated by the ring strain imposed by the bond angles of 60°, and thus, degradation in the presence of transition metals is very common [234]. Alternatively, ring insertion reactions may take place and are observed with main group element fragments such as $[SiR_2]$ to give $[R_2Si \subset P_4]$ (Fig. 10a) and $[(R_2Si)_2 \subset P_4]$ (Table 7).

In contrast to the rare examples of planar rings of tetrel atoms such as in $[(\eta^5\text{–}Pb_5)\{Mo(CO)_3\}_2]^{4-}$ (Fig. 9a), other planar Pn_x rings have been found in transition metal complexes. For Pn = P, the compounds $[(\eta^4\text{–}P_4)Nb(C_5Me_5)(CO)_2]$ (Fig. 10b), $[(\eta^4\text{–}P_4)Co(C_5Me_5)(\eta^2\text{–}P_2)\{Co(C_5Me_5)\}_2]$, and $[(\eta^1,\eta^4\text{–}P_4)\{W(CO)_5\}_4\{W(CO)_4\}]$ contain four-membered rings [239, 240], and the sandwich and triple-decker complexes $[(\eta^5\text{–}P_5)Fe(\eta^5\text{–}C_5Me_5)]$ and $[Cp^*Mo(\eta^6\text{–}P_6)MoCp^*]$ (Fig. 10d) [243], both of which have been obtained from reactions with white phosphorus, display five and six-membered rings, respectively. A formal replacement of all CH units in benzene by isoelectronic P atoms leads to the phosphaanalog P_6, and thus the latter compound can be regarded as a hexaphosphabenzene complex. According to the 18-electron rule, $[Cp^*Mo(\eta^6\text{–}P_6)MoCp^*]$ must contain a Mo–Mo single bond which is in agreement with a rather short Mo–Mo distance. The sandwich complex $[(\eta^5\text{–}P_5)_2Ti]^{2-}$ (Fig. 10c) displays two coplanar P_5 rings and can be described as a Ti^0 complex of the aromatic phosphacyclopentadienide in analogy to $(C_5H_5)^-$. Examples of the large number of transition metal P_n complexes with $n = 6$, 7, and 8 are listed in Table 7. P_6 units are often less symmetric, whereas P_8 complexes preferably show a symmetric realgar-type structure as depicted for $[(P_8)(SmCp^*{}_2)_4]$ in Fig. 10j, and further examples are stabilized as FeCp(CO) complexes [256].

Most frequent are P_7 complexes in which the heptaphosphanortricyclane unit is retained. In the protonated anion $[(P_7)Pt(H)PPh_3]^{2-}$ [247], the platinum atom coordinates side-on to two lone pairs of two divalent P atoms. When the related Ni(CO) complex shown in Fig. 10h is protonated, the incoming proton is attached

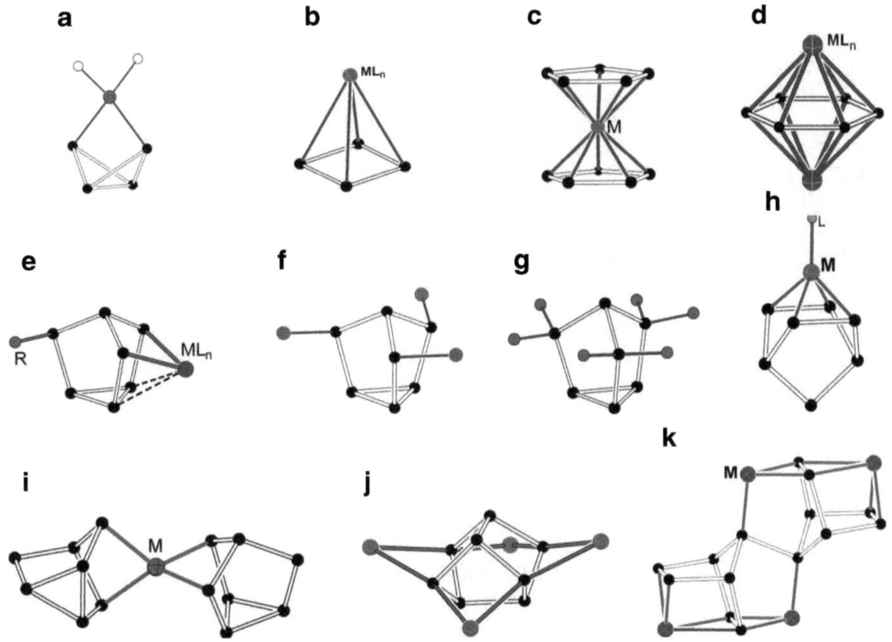

Fig. 10 Molecular structures of functionalized pnictide cages and pnictide complexes. a) P_4SiR_2, b) $(\eta^4–P_4)ML_n$, c) $[(\eta^5–P_5)_2Ti]^{2-}$, d) $[(\eta^6–P_6)(ML)_2]$, e) $[(P_7\,R)(ML_3)]^{2-}$, f) P_7R_3, g) $[P_7R_6]^{3+}$, h) $[(\eta^4–P_7)ML_n]^{3-}$, i) $[(P_7)_2M]^{4-}$, j) $[(P_8)(ML_2)_4]$, k) $[(P_{14})(ML_2)_4]$. Further examples are given in Table 7

to the remaining divalent P atom. A coordination with two further but longer M–P bonds is observed in the complex $[Nb(OC[^2Ad]Mes)_3(P_7PH_2)]$ of the mono-functionalized cage $[P_7–PH_2]^{2-}$ (Fig. 10e). $[(P_7)_2M]^{4-}$ (M = Zn, Cd) (Fig. 10i) contains two coordinated P_7 units, whereas in $[P_7\{FeCp(CO)_2\}_3]$ covalent M–P *exo*-bonds are formed (Fig. 10f). In $[P_7In(C_6H_5)_2]^{2-}$, the InR_2 group bridges two divalent P atoms of the P_7 cage with the In atom in a tetrahedral environment [228]. P_7 cages can also be formed during the reaction of white phosphorus by the insertion of P-containing species into the P_4 skeleton. Recently, a stepwise extension of the P_4 cage was achieved using $[PPh_2]^+$ [238], which led to the formation of $[Ph_2P \subset P_4]^+$, $[(Ph_2P)_2 \subset P_4]^{2+}$, and $[(Ph_2P)_3 \subset P_4]^{3+}$. The latter is shown in Fig. 10g and can formally be considered as the product of the addition of 6 R^+ fragments to a $P_7{}^{3-}$ unit. A neutral P_7Me_3 species has been obtained by reacting Li_3P_7 with Me_3X. This direct route has also been applied for the synthesis of hydrogenpolyphosphides and substituted phosphides [261], and can also be used to introduce main group organyls as substituents as in $P_7(MMe_3)_3$ (*M* = Si, Ge, Sn, Pb) [252].

The complex anions $[(\eta^4–P_7)Ni(CO)]^{3-}$ [247] and $[(\eta^4–P_7)M(CO)_3]^{3-}$ (M = Cr, Mo, W) shown in Fig. 10h are an exception, since their P_7 units cannot be

Table 7 List of structurally characterized phosphorus cage compounds

Formula		Ref.	Fig.
$(\eta^1\text{–}P_4)ML_3L'_2$	M = W; L = CO, L′ = PCy_3	[235]	
E ⊂ P_4, (E⊂)$_2$$P_4$	E = aminosilyl	[236]	10a
P_4R_2	R = Ar^{Dipp2} = C_6H_3-2,6-(C_6H_2-2,6-^{i}Pr_2)$_2$	[237]	
$(P_4R_2)M_2$	M = Nb, R = Ar^{Dipp2}	[237]	
$[P_5R_2]^+$ or $[(PR_2) \subset P_4]^+$	R = C_6H_5	[238]	
$(\eta^4\text{–}P_4)ML_n$	ML_n = $Nb(C_5Me_5)(CO)_2$	[239]	10b
$[(\eta^1,\eta^4\text{–}P_4)(ML_5)_4(ML_4)]$	M = W, L = CO	[240]	
$[P_4(ML)\ P_2(ML)_2]$	ML = Co(η^5–C_5Me_5)	[231]	
$[(P_4)_2\ M]^+$	M = Cu	[233],	
	M = Ag	[241]	
$[(P_5)_2M]^{2-}$	M = Ti	[242]	10c
$[(\eta^5\text{–}P_5)ML]$	M = Fe, L = η^5–C_5Me	[243]	
$[(\eta^6\text{–}P_6)(ML)_2]$	M = Mo, L = η^5–C_5Me_5	[244]	10d
$[(P_6)(ML)_3(P_3)M]$	M = Fe, L = Cp	[231]	
$[P_6(ML)_3]^+$	M = Fe, L = Cp	[231]	
	M = Nb, L = η^5–Cp′	[239]	
$[P_7R]^{2-}$	R = H	[245, 246]	
	R = $InPh_2$	[228]	
$[(P_7)MRL]^{2-}$	R = H, M = Pt, L = PPh_3	[247]	10e
$[(P_7R)ML_n]^{2-}$	R = H; M = Mo, W; L = CO, $n = 4$	[248]	10h
$[(P_7R)ML_3]$	R = PH_2, M = Nb, L = (OC)[^{2}Ad]Mes	[249]	10e
$[P_7R_2]^-$	R = H	[250]	
	R = CH_2Ph	[224]	
$[P_7RR']^{2-}$	R = $Si(SiMe_3)_3$, R′ = P–R	[251]	
P_7R_3	R = EMe_3, E = Si–Pb	[252]	10f
$[P_7(ML_n)_3]$	ML_n = $FeCp(CO)_2$	[231]	10f
$[P_7R_6]^{3+}$	R = Ph	[238]	10g
$[(\eta^4\text{–}P_7)ML_n]^{3-}$	M = Ni, L = CO, $n = 1$	[247]	10h
	M = Cr, Mo, W; L = CO, $n = 3$	[225, 253]	
$[(P_7)_2M]^{4-}$	M = Zn, Cd	[228]	10i
$[(P_8)(ML_2)_4]$	M = Sm, L = Cp*	[254]	10j
$[(P_8)(ML)_3]$	M = Co, L = ^{t}Bu_3Cp	[255]	
$[(P_8)(MLL')_2(MLL'_2)_2]$	M = Fe, L = Cp′, L′ = CO	[256]	
$[(P_8)\ (MLL')_3(MLL'_2)$ $(ML'_4)_2]$	M = Fe, L = Cp′, L′ = CO	[256]	
$[(P_8)(MLL')_2(M'L_5)_3]$	M = Ir, M′ = Cr, L = Cp*, L′ = CO	[257]	
$[P_{11}R]^{2-}$	R = H	[127, 258, 259]	
$[(P_{12})(ML)_3]$	R = ML; M = Co, L = ^{t}Bu_3Cp	[255]	
$(NHC)_2(P_{12})$	(NHC = $C(NDipp)_2C_2H_4$, Dipp = 2,6-^{i}PrC_6H_3)	[260]	
$[(P_{14})(ML_2)_4]$	M = Ni, L = PBu_3	[231]	10k

described with localized bonds. Their structures can only be explained by including π-type orbitals of the P_7 cage that act as a 4-electron-donor to a 14-electron $[Cr(CO)_3]^{2-}$ fragment [225].

In $[(Ni(PBu_3)_2)_4P_{14}]$ (Fig. 10k), two $[P_7]^{3-}$ clusters are coupled under reduction of the polyanion. The resulting $[P_{14}]^{8-}$ formally consists of two norbornane-like units covalently linked via their apical P atoms, and follows the 8-N rule.

4.2 Arsenic

In contrast to the P-containing clusters, the corresponding As cages have a larger tendency to form intermetalloid cluster (Table 8). The reaction of $[AsCH_3]_5$ with $Co_2(CO)_8$ in hexane leads to the formation of *cyclo*-As_3 [222] and *cyclo*-As_5 units. According to the isolobal concept, the resulting complex $[As_3(ML_3)]$ (Fig. 11a) is electronically equivalent to the As_4 molecule [223]. In the analogous reaction of $[AsCH_3]_5$ with $[CpMo(CO)]_2$ in toluene, the triple-decker complex $[(\eta^5–C_5Me_5)Co(\eta^5–As_5)Co(\eta^5–C_5Me_5)]$ with a planar As_5 ring is obtained (Fig. 11b). Because of the observed bond length variation within the As_5 ring and of one longer Mo–As contact, a $[(\eta^2–As_3)]/[(\eta^2–As_2)]$ coordination mode seems to be reasonable, and – as in the P_6 triple-decker complex discussed above – the M–M distance is again rather short [223]. $[NbAs_8]^{3-}$ [221] and $[MoAs_8]^{2-}$ [230] contain S_8-analogous eight-membered $[As_8]^{8-}$ rings coordinated to Nb(V) and Mo(VI) cations, respectively (Fig. 11d). Eight two-electron σ-donor atoms per ring lead to 16 electron complexes. For their stabilization in terms of an 18 electron species, additional π-donation from the $[As_8]^{8-}$ unit to the central metal has been suggested.

As observed for the corresponding phosphorus cluster, the $[As_7]^{3-}$ cage can accommodate an $M(CO)_3$ fragment in an η^4-fashion (Fig. 11c), and the protonated species $[(As_7H)]^{2-}$ binds the $PtPPh_3$ moiety in an η^2-coordination mode (Fig. 10e) [227]. Dimeric complexes of $[As_7]^{3-}$ with the late transition metals Pd and Cu contain M_2 dumbbells and have the composition $[As_7(M_2)As_7]^{4-}$. For M = Cu, the nortricyclane cages remains intact, and thus a description of two $[As_7]^{3-}$ ligands that are attached to a $Cu_2{}^{2+}$ dimer is appropriate (Fig. 11e) [228]. However, for M = Pd (Fig. 11f), one bond of the triangle of each As_7 unit is opened to a norbornane type and just in analogy to the ML_3 complex shown in Fig. 11c. Assuming five divalent As^{1-} atoms within a $[As_7]^{5-}$ ligand, a $[Pd_2]^{6+}$ dumbbell is required for charge compensation of two ligands. Dinuclear Pt(III) complexes with Pt–Pt bonds are common, but Pd(III) dimers are not, and therefore the electronic situation in this complex is probably more complex.

Table 8 List of structurally characterized arsenic cage compounds

Formula		Ref.	Fig.
$[As_3(ML_3)]$	M = Co, L = CO	[222]	11a
$[(As_5)(ML)_2]$	ML = Co, L = η^5–C_5Me_5	[223]	11b
$[As_7R_2]^+$	R = CH_2Ph	[224]	
$[(\eta^4–As_7)ML_3]^{3-}$	M = Cr, Mo, W; L = CO	[225, 226]	11c
$[(As_7)MHL]^{2-}$	M = Pt, L = PPh_3	[227]	10e
$[(As_7)_2(M_2)]^{4-}$	M = Cu	[228]	11e
	M = Pd	[229]	11f
$[(As_7)_2 M]^{4-}$	M = Sn	[135]	11g
$[As_8M]^{n-}$	M = Nb, n = 3	[221]	11d
	M = Mo, n = 2	[230]	11d
$[(As_6)(As_3)_2(ML)_6]$	M = Co, L = PEt_2Ph	[231]	11h
$[As_{16} M_7]^{4-}$	M = Pd	[229]	11i
$[As@M_{12}@As_{20}]^{3-}$	M = Ni	[17]	11j

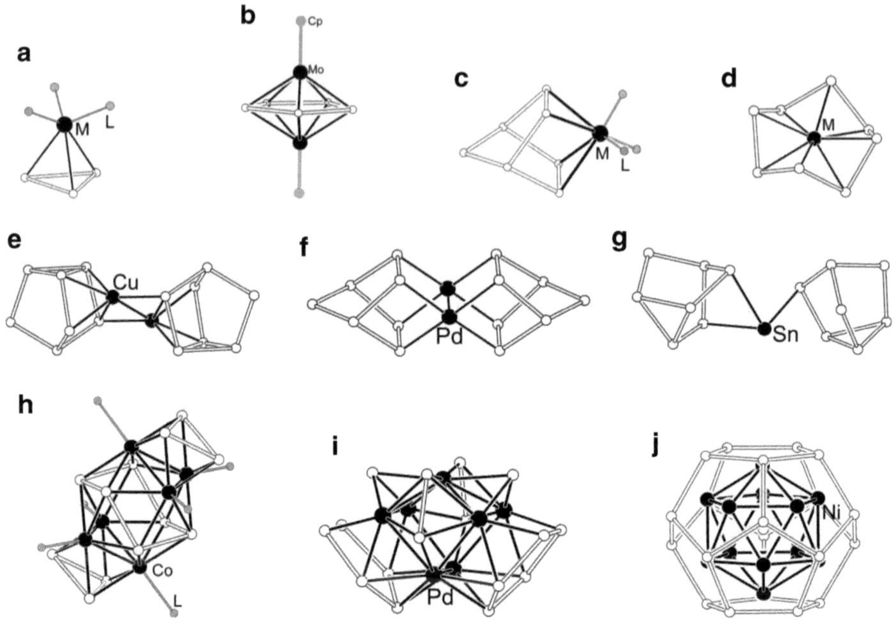

Fig. 11 Structures of arsenide cage complexes and intermetalloid clusters. a) $[(\eta^3\text{–}As_3)(ML_3)]$, b) $[(\eta^5\text{–}As_5)(ML)_2]$, c) $[(\eta^4\text{–}As_7)ML_3]^{3-}$, d) $[As_8M]^{n-}$, e) $[(As_7)_2(Cu_2)]^{4-}$, f) $[(As_7)_2(Pd_2)]^{4-}$, g) $[(As_7)_2 Sn]^{4-}$, h) $[(As_6)(As_3)_2(ML)_6]$, i) $[As_{16} M_7]^{4-}$, j) $[As@Ni_{12}@As_{20}]^{3-}$. Further examples are given in Table 8

Dissolution of solid KSnAs in appropriate solvents leads to the formation of main group metal-connected As_7 units of the type $[As_7SnAs_7]^{4-}$ (Fig. 11g). In order to achieve a charge balance, the tin atoms should have a formal oxidation state of +2, in analogy to the Zn and Cd atoms in the P_7 complexes discussed above. The Sn–As bonds in $[As_7SnAs_7]^{4-}$ are significantly longer than those observed for the trivalent Sn atoms in KSnAs, but three shorter bonds and one long nonbonding Sn–As contact can clearly be distinguished.

The reaction of K and As in the ration 3:7 leads to products which in the presence of Co(II) fragments in liquid ammonia give a complex depicted in Fig. 11h, which contains As_3 and As_6 rings. Two Co_3 triangles built by three $Co(PEt_2Ph)$ units are attached to the As_6 ring with chair conformation in such a way that a distorted As_6Co_6 icosahedron is formed. The two Co triangles are additionally coordinated to two As triangles which leads to two distorted As_3Co_3 octahedra, each of which is face-sharing via its Co_3 triangular face with the icosahedron. Assuming solely As^{1-} ions, which according to the 8-N rule form chalcogene-type $[As_3]^{3-}$ and $[As_6]^{6-}$ rings, for the As substructure a total charge of 12− results, and this charge is balanced by six Co(II) ions in $[(As_6)(As_3)_2\{Co(PEt_2Ph)\}_6]$. In a similar way, a complete rearrangement of the $[As_7]^{3-}$ anion in K_3As_7 takes place when it is reacted with $Pd(PCy_3)_2$. The Pd-rich anion $[Pd_7As_{16}]^{4-}$ in Fig. 11i contains a Pd_7 core that coordinates to two planar As_5 rings, two As_2 units, and two As atoms.

Assuming as above two $[As_5]^{1-}$ rings with a 6π electron system, two $[As_2]^{2-}$ dumbbells, and two isolated As^{3-} atoms, the resulting $12 - 4 = 8$ negative charges of the As skeleton are counterbalanced by six Pd(I) and one Pd(II) counterions. The latter one is surrounded by four As atoms in a square-planar arrangement [229]. The stuffed intermetalloid cluster $[As@Ni_{12}@As_{20}]^{3-}$ (Fig. 11j) contains condensed five-membered As rings, which form a As_{20} pentagondodecahedron and a central As atom which is icosahedrally surrounded by 12 Ni atoms. The resulting endohedral icosahedron $[Ni_{12}(\mu_{12}\text{–}As)]$ is embedded in the As_{20} pentagondodecahedron with exclusively trivalent As atoms. In a simple picture, all atoms of the As_{20} pentagondodecahedron and the endohedral As^{3-} atom are electronically saturated when a $Ni(0)_{12}$ icosahedron is assumed. A more complex analysis of the electronic structure is given elsewhere [17].

4.3 Antimony and Bismuth

For the pnictides, Sb and Bi transition metal-rich complexes and intermetalloid clusters are clearly favored (Table 9). In $[Sb_3Ni_4(CO)_6]^{3-}$ (Fig. 12a), a bent Sb_3 unit is coordinated to four Ni atoms, and an aromatic $C_5H_5^{-}$-analogous Sb_5 ring is found in $[Cp'Mo(\eta^5\text{–}Sb_5)MoCp']$ (Fig. 12b). By shape and electron count, the Sb_3Ni_4 deltahedron of the former anion corresponds to a 16 ske *closo* cluster. No transition metal complex of an intact nortricyclane is known as yet, but the three-membered ring of this unit is opened upon complexation in $[(\eta^4\text{–}Sb_7)NiCO]^{3-}$ and $[(\eta^4\text{–}Sb_7)M(CO)_3]^{3-}$, and the resulting structures are similar to those discussed above and shown in Figs. 10h and 11c. An $[Sb_7]^{3-}$ polyanion with an alternative structure different from nortricyclane is formed upon coordination of three Ni(CO)

Table 9 List of structurally characterized cage compounds of antimony and bismuth

Formula		Ref.	Fig.
$[Sb_3M_4L_6]^{3-}$	M = Ni, L = CO	[262]	12a
$[Sb_3(MLL'_2)]$	M = Mo, L = Cp/Cp*, L′ = CO	[263]	
$[Sb_4(ML_2)_4]^{2+}$	M = Pd, L = PPh_2Me	[264]	12g
$[(\eta^5\text{–}Sb_5)(ML)_2]$	M = Mo; L = $C_5H_2R_3$, R = Me, ^{t}Bu	[265]	12b
$[(\eta^4\text{–}Sb_7)ML_3]^{3-}$	M = Cr, W; L = CO	[225]	11c
	M = Mo, L = CO	[266]	11c
$[(Sb_7)(ML)_3]^{3-}$	M = Ni, L = CO	[267]	12c
$[Sb_{17}M_5]^{4-}$	M = Ni	[268]	12d
$[Bi_3(ML_3)_2]^{3-}$	M = Cr, Mo; L = CO	[269]	12e
$[Bi_3M_4L_6]^{3-}$	M = Ni, L = CO	[262]	12a
$[Bi_3M_6L_9]^{3-}$	M = Ni, L = CO	[262]	12f
$[Bi_4(ML_3)_3(ML_4)]^{2-}$	M = Fe, L = CO	[270]	
$[Bi_4(ML_3)_3(FeL_2L')]$	M = Fe, L = CO, L′ = Cp′	[271]	
$[Bi_4(ML_2)_4]^{2+}$	M = Pd, L = PPh_2Me	[264]	12g
$[Bi_4M_4L_6]^{2-}$	M = Ni, L = CO	[262]	12h
$[M_x@Bi_6M_6L_8]^{4-}$	M = Ni, L = CO	[262]	12i
$[M@M_8Bi_4@Bi_7]^{5-}$	M = Zn	[272]	12j
$[M@M_5Sn_3Bi_3@Bi_5]^{4-}$	M = Zn	[273]	

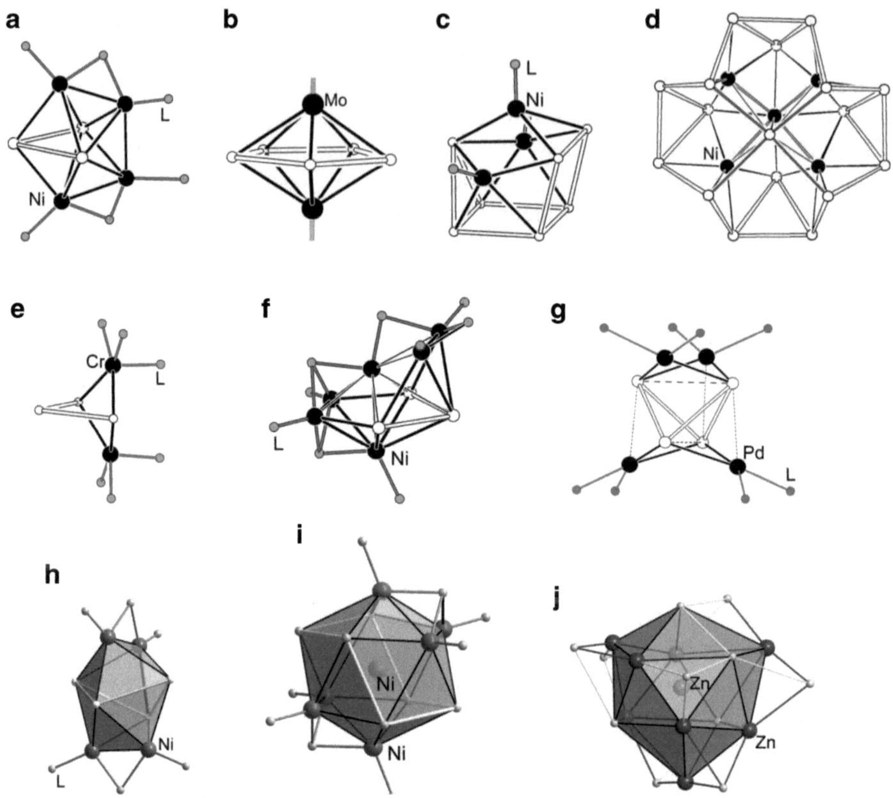

Fig. 12 Molecular structures of functionalized antimonide (a–d) and bismuthide (e–j) cages. a) $[Sb_3M_4L_6]^{3-}$, b) $[(\eta^5\text{–}Sb_5)(ML)_2]$, c) $[(Sb_7)(ML)_3]^{3-}$, d) $[Sb_{17}M_5]^{4-}$, e) $[Bi_3(ML_3)_2]^{3-}$, f) $[Bi_3M_6L_9]^{3-}$, g) $[Bi_4(ML_2)_4]^{2+}$, h) $[Bi_4M_4L_6]^{2-}$, i) $[M_x@Bi_6M_6L_8]^{4-}$, j) $[M@M_8Bi_4@Bi_7]^{5-}$. Further examples are given in Table 9

fragments. In $[(Sb_7)\{Ni(CO)\}_3]^{3-}$ (Fig. 12c), the three- and four-membered rings are connected in a cuneane-like manner, and the anion contains three Sb atoms with two homoatomic bonds. Thus, the charge of the $[(Sb_7)\{Ni(CO)\}_3]^{3-}$ unit is localized on the atoms of the Sb polyanion. According to Wade's rules this cluster can also be regarded as a 24 ske *nido*-type cluster ($2 \times 10 + 4 = 24$). However, the *nido*-type differs from the other known nido complexes (e.g., $B_{10}H_{14}$) that have open six-membered rings, since it contains two four-membered rings fused along a common edge. Thus, again the description of an intermediate structure between a deltahedron and a cube is more appropriate [267].

An early stage of the consecutive connection of five-membered rings to reach the perfect As_{20} cage of Fig. 11i is trapped in the intermetalloid cluster $[Sb_{17}Ni_5]^{4-}$ (Fig. 12d). It consists of a covalent Sb_{16} crown-shaped polyanion which is connected via homoatomic bonds to the central Sb atom (gray bonds), with, however, longer distances than the other Sb–Sb bonds. The central Ni_5 unit has a similar shape as five of the seven Pd atoms in $[Pd_7As_{16}]^{4-}$ shown in Fig. 11i. In this case, a description of the bonding situation including delocalized electronic states of the 139 valence electrons is probably more appropriate. An EPR signal of this paramagnetic anion has not been obtained yet [268].

For Pn = Bi also smaller polyanionic units are known. One example is the rather simple $[(\eta^2\text{–}Bi_3)(\mu^2\text{–}M)_2(CO)_6]^{3-}$ anion (M = Cr, Mo) [269] which contains an ozone-like $[Bi_3]^{3-}$ unit (Fig. 12e; Table 9). A trigonal-pyramidally distorted Bi_4 tetrahedron is formed in $[Fe_4(Bi_4)(CO)_{13}]^{2-}$ which can be obtained from $[BiFe_3(CO)_{12}]^-$ precursors. In $[Fe_4(Bi_4)(CO)_{13}]^{2-}$, three faces of the Bi_4 unit are capped by $Fe(CO)_3$ moieties, and an $Fe(CO)_5$ unit is attached to the apical Bi atom, which has longer bonds to the Bi atoms of the empty face [270]. An even more distorted Bi_4 unit occurs in $[Bi_4\{Pd(PPh_2Me)_2\}_4]^{2+}$ (Fig. 12g) as it was also observed for the Sb analog. It contains a $[Bi_4]^{6-}$ anion, if an oxidation state +2 is assumed for Pd. Two Bi–Bi bonds of the Bi_4 tetrahedron are elongated, but are still bonding in character. A delocalized bonding situation can also be attributed to another distinctly longer Pd–Bi contact (dashed lines).

The reaction of K_4Bi_5 with $Ni(PPh_3)_2(CO)_2$ in ethylenediamine leads to another series of heteroatomic cluster compounds [262]. In the deltahedral *closo* clusters $[Bi_3Ni_4(CO)_6]^{3-}$ (Sb analog in Fig. 12a), $[Bi_4Ni_4(CO)_6]^{2-}$ (Fig. 12h), and $[Ni_x@Bi_6Ni_6(CO)_8]^{4-}$ (Fig. 12i), three- and four-atomic fragments occur and with 16, 18, and 26 ske, respectively. In all cases Bi contributes three and the Ni–CO fragments zero electrons to the cluster bonding. Additional CO ligands supply two additional electrons, and also the overall charges of the clusters have to be taken into account. The site at the center of the icosahedron in $[Ni_x@Bi_6Ni_6(CO)_8]^{4-}$ is only partially occupied with Ni, probably as a consequence of the fact that the central Ni^0 atom is not required for the cluster bonding, as it is also the case for the empty and filled clusters $[Pb_{10}]^{2-}$ and $[Ni@Pb_{10}]^{2-}$, respectively (see above). The structure of the Bi_6 unit within an icosahedron in Fig. 12i (represented with gray bonds) shows a striking similarity to that of the icosahedron with the As_6 substructure in Fig. 11h. Since in the latter the metal atoms are further connected to two As_3 units, Wade's rules do not apply. A transition metal-rich variant attached to a Bi_3 unit is realized in $[Bi_3Ni_6(CO)_9]^{3-}$ (Fig. 12f), which from a structural point of view appears to be a fragment of the $[Ni_x@Bi_6Ni_6(CO)_8]^{4-}$ icosahedron shown in Fig. 12i, which results when a Bi_3Ni unit is removed from the filled icosahedron ($x = 1$). Both structures also contain a similar μ^3-capping CO ligand.

An Zn-centered Zn_8Bi_4 icosahedron appears in the endohedral cluster anion $[Zn@(Zn_8Bi_4)Bi_7]^{5-}$ (Fig. 12j). In contrast to the situation in Fig. 12i, it shows only one Bi–Bi contact along the edges of the icosahedron. Further bonds appear to seven Bi atoms in capping positions. Molecular orbital considerations reveal a delocalized bonding situation with a distinct HOMO-LUMO gap; however, there is no straightforward way to use Wade's ske counting rules. Zn-centering also occurs in the recently discovered *ternary* intermetalloid cluster anion $[Zn@Zn_5Sn_3Bi_3@Bi_5]^{4-}$. Since the structure suffers from strong disorder of the Sn/Bi atom positions, an exact determination of the composition is difficult.

However, a quantum-chemical analysis indicates the stability of the $Zn_6Sn_3Bi_8$ cluster core. It can be described as a Zn-centered monocapped pentagonal antiprism, which is equivalent to an icosahedron with one missing apex atom, and consists of a planar Zn_5 unit which is coplanar with an $(Sn/Bi)_5$ five-membered ring. As in $[Zn@(Zn_8Bi_4)Bi_7]^{5-}$ (Fig. 12j), five Bi atoms act as caps on the deltahedral faces [273].

5 Summary and Concluding Remarks

The numerous examples of structures of group 14 and 15 element cluster anions and molecules allow the following conclusions:

(a) There is a strong structural similarity between soluble Zintl ions and their reaction products, intermetalloid clusters, and cage molecules. The formal addition of ligands to preformed polyanions (Zintl clusters) leads to structures which in the case of tetrel elements are mainly synthesized from monomeric units with low-valent tetrel atoms, and some of them are classified as intermetalloid clusters (Table 10).
Furthermore, evident structural similarities between compounds obtained through reaction with Zintl ion clusters and ligand-stabilized cage molecules occur (Table 11).

(b) When heteroatomic clusters reach approximately the size of ten vertex atoms, covalent bonding and Wade's electron-counting rules are increasingly violated. Thus, *intermetalloid clusters* or molecular alloys with shapes that deviate from deltahedral structures appear. Open four- and five-membered rings in non-deltahedral topologies are formed:

For Ge in $[Ge_{10}(Si^tBu_3)_6I]^+$ (Fig. 5p), $[Ge_{10}(SiR_3)_4(SiR_2)R']^-$ (Table 4), $[Ge_{14}\{Ge(SiMe_3)_3\}_5]^{3-}$ (Fig. 5q), $[Co@Ge_{10}]^{2-}$ (Fig. 6i), $[Ge_{45}Au_3]^{9-}$ (Fig. 6c)
For Sn in $[Sn_4\{Sn(C_6H_3\text{-}2,6\text{-}(2,4,6\text{-}Me_3C_6H_2)_2)\}_4]$ (Fig. 7d) and $[Sn_4\{Sn(Si^tBu_3)\}_4]^{2-}$ (Fig. 7e) and $[Sn_4\{Sn(Si^tBu_3)\}_4]$ (Table 5) and $[Sn@\{Sn_8(SnR)_6\}]$ (R = $N(2,6\text{-}^iPr_2C_6H_3)(SiMe_2R')$; R′ = Me, Ph)
For As in $[(As_5)_2\ (As_2)_2\ As_2\ Pd_7]^{4-}$ (Fig. 11i) and $[As@Ni_{12}@As_{20}]^{3-}$ (Fig. 11j); and for Sb in $[(Sb_7)\{Ni(CO)\}_3]^{3-}$ (Fig. 12c)

(c) Most interestingly, also hybrids between deltahedral and non-deltahedral structures are common such as $[Ge_{10}\{Fe(CO)_4\}_8]^{6-}$ (Fig. 5o), $[M@Ge_9\text{–}M'L]^{2/3-}$ (Fig. 6e; Table 4), $[Ni_3@Ge_{18}]^{4-}$ (Fig. 6f), $[M_2@Sn_{17}]^{4-}$ (Fig. 8g), $[Pd_2@Sn_{18}]^{4-}$ (Fig. 8h), $[Sn_{17}\{GaCl(ddp)\}_4]$ (Fig. 8j), and $[Pb_{10}\{Si(SiMe_3)_3\}_6]$ (Fig. 9g).

In summary, the bonding situation found in the deltahedral structures of bare Zintl ions, which follow Wade's rules, changes smoothly to that observed in ligand-stabilized metalloid and heteroatomic clusters. Therefore, they can be considered as a subclass of intermetalloid clusters. The accessibility of larger intermetalloid

Table 10 Relation of Zintl ions and ligand-stabilized cage compounds

$[E_n]^{x-} + m\,R^+$		$\rightarrow[E_nR_m]^{(x-m)-}$	
$[Si_4]^{4-}$	(Fig. 1)	$[Si_4R_3]^-$	(Fig. 2b)
$[Si_4]^{4-}$	(Fig. 1)	$[Si_4R_4]$	(Fig. 2a)
$[Ge_6]^{2-}$ [a]	(Fig. 7b)	$[Ge_6R_2]$	(Fig. 5b)
$[Ge_9]^{4-}$	(Fig. 2d)	$[Ge_9R]^{3-}$	(Fig. 5d)
$[Ge_9]^{4-}$	(Fig. 2d)	$[Ge_9R_2]^{2-}$	(Fig. 5e)
$[Ge_9]^{4-}$	(Fig. 2d)	$[Ge_9R_3]^-$	(Fig. 5f)
$[(\eta^5\text{–}E_9)M(CO)_3]^{4-}$ [b]	(Figs. 7j, 9e)	$[(\eta^5\text{–}Ge_9R_3)Cr(CO)_3]^-$	(Fig. 5h)
$[Sn_9]^{3-}$	(Fig. 2c)	$[Sn_9R_3\bullet]$	(Fig. 7g)
$[Sn_{10}]^{2-}$ [c]	(Fig. 2e)	$[Sn_{10}R_3]^+$	(Fig. 7l)
$[Ni_2@Sn_{17}]^{4-}$	(Fig. 8g)	$[\square_2@Sn_{17}R_4]$	(Fig. 8j)
$[P_7]^{3-}$	(Fig. 3a)	$[P_7R_3]$	(Fig. 9f)
$[As_7]^{3-}$	(Fig. 3a)	$[As_7R_2]^+$ [d]	(not shown)
$[P_7]^{3-}$	(Fig. 3a)	$[P_7R_6]^{3+}$	(Fig. 9g)

[a] Unit in, e.g., $[Ge_6\{Cr(CO)_5\}_6]^{2-}$ (Table 4)
[b] Hypothetical, known for E = Sn and Pb only
[c] Hypothetical, known for E = Ge and Pb only
[d] after oxidation

Table 11 Relation of Zintl ion clusters and cage molecules

Zintl ion clusters		Cage molecule	
$[M@Pb_{12}]^{2-}$	(Fig. 9h)	$[\square@Pb_{12}R_6]$	(Fig. 9i)
$[\eta^5\text{–}Pb_5]\{Mo(CO)_3\}^{4-}$	(Fig. 9a)	$[(\eta^5\text{–}As_5)\{CoCp^*\}_2]$	(Fig. 11b)
		$[(\eta^5\text{–}Sb_5)\{Mo(C_5H_4R)\}_2]$	(Fig. 12b)
		$(\eta^5\text{–}Sn_5)\{Sn(2,6\text{-}(2,6\text{-}^iPr_2C_6H_3)_2C_6H_3)\}_2$	(Fig. 7c)

clusters via deltahedral homoatomic Zintl ions may be very useful for syntheses in the field of materials science, and finally, the recent discovery of the first silicon-containing Zintl ions that can serve as building blocks for larger cluster units has provided another promising aspect, since silicon-based materials will have an even larger impact than, e.g., germanium-based materials.

Acknowledgments The author thanks the Deutsche Forschungsgemeinschaft (DFG) for continuous financial support. Special thanks go to S. Scharfe, M. Waibel, F. Kraus, and V. Hlukhyy for their support to create the figures and A. Schier for her effort to improve this manuscript.

References

1. Irngartinger H, Goldman A, Jahn R, Nixdorf M, Rodewald H et al (1984) Angew Chem Int Ed Engl 23:993–994
2. Troyanov SI, Popov AA (2005) Angew Chem Int Ed 44:4215–4218
3. Wiberg N, Finger CMM, Polborn K (1993) Angew Chem Int Ed Engl 32:1054–1056
4. Klinkhammer KW, Xiong Y, Yao S (2004) Angew Chem Int Ed 43:6202–6204
5. Ecker A, Weckert E, Schnöckel H (1997) Nature 387:379
6. Schnepf A, Schnöckel H (2001) Angew Chem Int Ed 40:712
7. Schnepf A, Schnöckel H (2002) Angew Chem Int Ed 41:3533–3552
8. Zintl E (1929) Naturwiss 17:782–783

9. Zintl E, Dullenkopf W (1932) Z Phys Chem B16:195
10. Schäfer R, Klemm W (1961) Z Anorg Allg Chem 312:214
11. Krätschmer W, Lamb LD, Fostiropoulos K, Huffman DR (1990) Nature 347:354–358
12. Schnöckel H (1996) Angew Chem Int Ed 35:129–149
13. Spiekermann A, Hoffmann SD, Fässler TF (2006) Angew Chem Int Ed 45:3459–3462
14. Wang J-Q, Stegmaier S, Wahl B, Fässler TF (2010) Chem Eur J 16:1793–1798
15. Fässler TF, Hoffmann SD (2004) Angew Chem Int Ed 43:6242–6247
16. Esenturk EN, Fettinger J, Eichhorn BW (2006) Polyhedron 25:521
17. Moses MJ, Fettinger JC, Eichhorn BW (2003) Science 300:778–780
18. Corbett JD (1985) Chem Rev 85:383–397
19. Corbett JD (1997) Struct Bond 87:158–193
20. Fässler TF (2001) Coord Chem Rev 215:347–377
21. Scharfe S, Fässler TF (2010) Phil Trans R Soc A 328:1265–1284
22. von Schnering H-G (1981) Angew Chem Int Ed Engl 20:33–51
23. von Schnering H-G, Hönle W (1988) Chem Rev 88:243–273
24. Scharfe S, Kraus F, Stegmaier S, Schier A, Fässler TF (2011) Angew Chem Int Ed (in press)
25. Schnöckel H (2010) Chem Rev 110:4125–4163
26. Tran NT, Dahl LF (2003) Angew Chem Int Ed 42:3533–3537
27. Fässler TF, Kronseder C, Wörle M (1999) Z Anorg Allg Chem 625:15
28. Zintl E (1939) Angew Chem 52:1–6
29. Klemm W (1958) Proc Chem Soc London:329
30. Busmann E (1961) Z Anorg Allg Chem 313:90–106
31. Wade K (1976) Adv Inorg Chem Radiochem 18:1–66
32. Zintl E, Harder A (1931) Z Phys Chem A 154:47
33. Kummer D, Diehl L (1970) Angew Chem Int Ed Engl 9:895
34. Smyth FH (1917) J Am Chem Soc 39:1299
35. Nienhaus A, Hoffmann SD, Fässler TF (2006) Z Anorg Allg Chem 632:1752–1758
36. Corbett JD, Adolphson DG, Merryman DJ, Edwards PA, Armatis FJ (1975) J Am Chem Soc 97:6267
37. Waibel M, Kraus F, Wahl B, Scharfe S, Fässler TF (2010) Angew Chem Int Ed 49:6611–6615
38. Marsh RE, Shoemaker DP (1953) Acta Cryst 6:197–205
39. Wiesler K, Brandl K, Fleischmann A, Korber N (2009) Z Anorg Allg Chem 635:508–512
40. Joseph S, Suchentrunk C, Kraus F, Korber N (2009) Eur J Inorg Chem:4641–4647
41. Hoch C, Wendorff M, Röhr C (2003) J Alloy Compd 361:206–221
42. Quéneau V, Todorov E, Sevov SC (1998) J Am Chem Soc 120:3263
43. Ponou S, Fässler TF (2007) Z Anorg Allg Chem 633:393–397
44. Quéneau V, Sevov SC (1997) Angew Chem Int Ed Engl 36:1754–1756
45. von Schnering H-G, Baitinger M, Bolle U, Carrillo-Cabrera W, Curda J et al (1997) Z Anorg Allg Chem 623:1037–1039
46. Kim S-J (2007) Substitution effects in binary intermetallic compounds: investigations in the system alkali and alkaline-earth metal – tin and alkaline-earth metal – bismuth. PhD Thesis, Technische Universität München, München
47. Hoch C, Wendorff M, Rohr C (2002) Acta Crystallogr C 58:I45–I46
48. Hoch C, Wendorff M, Röhr C (2003) Z Anorg Allg Chem 629:2391–2397
49. Quéneau V, Sevov SC (1998) Inorg Chem 37:1358–1360
50. Todorov E, Sevov SC (1998) Inorg Chem 37:3889–3891
51. Manriquez V, Hönle W, von Schnering H-G (1986) Z Anorg Allg Chem 539:95–109
52. Meyer T, Hönle W, von Schnering H-G (1987) Z Anorg Allg Chem 552:81–89
53. Hönle W, Buresch J, Peters K, Chang JH, von Schnering H-G (2002) Z Kristallogr NCS 217:485–486
54. Emmerling F, Röhr C (2002) Z Naturforsch 57b:963–975

55. Hönle W, Buresch J, Wolf J, Peters D, Chang JH, von Schnering H-G (2002) Z Kristallogr NCS 217:489–490
56. Somer M, Hönle W, von Schnering H-G (1989) Z Naturforsch 44b:296–306
57. von Schnering H-G, Somer M, Kliche G, Hönle W, Meyer T et al (1991) Z Anorg Allg Chem 601:13–30
58. Emmerling F, Röhr C (2003) Z Anorg Allg Chem 629:467–472
59. Hirschle C, Röhr C (2000) Z Anorg Allg Chem 626:1992–1998
60. Zintl E, Kaiser H (1933) Z Anorg Allg Chem 211:113
61. Eisenmann B (1993) Angew Chem Int Ed Engl 32:1693–1695
62. Zintl E, Dullenkopf W (1932) Z Phys Chem B 16:183
63. Fässler TF, Hoffmann R (1999) Angew Chem Int Ed 38:543–546
64. Goicoechea JM, Sevov SC (2004) J Am Chem Soc 126:6860–6861
65. Campbell J, Schrobilgen GJ (1997) Inorg Chem 36:4078–4081
66. Suchentrunk C, Korber N (2006) New J Chem 30:1737–1739
67. Edwards PA, Corbett JD (1977) Inorg Chem 16:903–907
68. Corbett JD, Edwards PA (1975) J Chem Soc Chem Commun:984–985
69. Goicoechea JM, Sevov SC (2005) Inorg Chem 44:2654–2658
70. Angilella V, Belin C (1991) J Chem Soc Faraday Trans 87:203–204
71. Fässler TF, Hunziker M (1994) Inorg Chem 33:5380–5381
72. Fässler TF, Schütz U (1999) Inorg Chem 38:1866
73. Critchlow SC, Corbett JD (1983) J Am Chem Soc 105:5715
74. Fässler TF, Hunziker M (1996) Z Anorg Allg Chem 622:837
75. Fässler TF, Hoffmann R (2000) Z Kristallogr NCS 215:139
76. Yong L, Hoffmann SD, Fässler TF (2005) Z Kristallogr NCS 220:49–52
77. Campbell J, Dixon DA, Mercier HPA, Schrobilgen GJ (1995) Inorg Chem 34:5798
78. Belin CHE, Corbett JD, Cisar A (1977) J Am Chem Soc 99:7163
79. Somer M, Carrillo-Cabrera W, Peters EM, Peters K, von Schnering H-G (1998) Z Anorg Allg Chem 624:1915–1921
80. Downie C, Mao J-G, Guloy AM (2001) Inorg Chem 40:4721
81. Suchentrunk C, Daniels J, Somer M, Carrilo-Cabrera W, Korber N (2005) Z Naturforsch 60b:277
82. Carrillo-Cabrera W, Aydemir U, Somer M, Kircali A, Fässler TF, Hoffmann SD (2007) Z Anorg Allg Chem 633:1575
83. Diehl L, Khodadadeh K, Kummer D, Strähle J (1976) Chem Ber 109:3404–3418
84. Corbett JD, Edwards PA (1977) J Am Chem Soc 99:3313–3317
85. Burns R, Corbett JD (1985) Inorg Chem 24:1489
86. Korber N, Fleischmann A (2001) Dalton Trans:383
87. Hauptmann R, Fässler TF (2002) Z Anorg Allg Chem 628:1500–1504
88. Hauptmann R, Fässler TF (2003) Z Kristallogr NCS 218:458
89. Hauptmann R, Fässler TF (2003) Z Kristallogr NCS 218:455–457
90. Hauptmann R, Hoffmann R, Fässler TF (2001) Z Anorg Allg Chem 627:2220–2224
91. Fässler TF, Hoffmann R (1999) J Chem Soc Dalton Trans:3339–3340
92. Yong L, Hoffmann SD, Fässler TF (2006) Inorg Chim Acta 359:4774–4778
93. Xu L, Sevov SC (1999) J Am Chem Soc 121:9245–9246
94. Hauptmann R, Fässler TF (2003) Z Kristallogr NCS 218:461–463
95. Hauptmann R, Fässler TF (2003) Z Anorg Allg Chem 629:2266
96. Ugrinov A, Sevov SC (2002) J Am Chem Soc 124:10990–10991
97. Yong L, Hoffmann SD, Fässler TF (2005) Z Anorg Allg Chem 631:1149–1153
98. Ugrinov A, Sevov SC (2003) Inorg Chem 42:5789–5791
99. Yong L, Hoffmann SD, Fässler TF (2004) Z Anorg Allg Chem 630:1977–1981
100. Downie C, Tang Z, Guloy AM (2000) Angew Chem Int Ed 39:338–340
101. Downie C, Mao J-G, Parmar H, Guloy AM (2004) Inorg Chem 43:1992
102. Belin C, Mercier H, Angilella V (1991) New J Chem 15:931

103. Kraus F, Korber N (2005) Chem Eur J 11:5945–5959
104. Kraus F, Aschenbrenner JC, Korber N (2003) Angew Chem Int Ed 42:4030–4033
105. Kraus F, Hanauer T, Korber N (2006) Inorg Chem 45:1117–1123
106. Hanauer T, Kraus F, Korber N (2005) Monatsh Chem 136:119–125
107. Korber N, Reil M (2002) Chem Commun:84–85
108. Critchlow SC, Corbett JD (1984) Inorg Chem 23:770–774
109. Cisar A, Corbett JD (1977) Inorg Chem 16:2482–2487
110. Kuznetsov AE, Fässler TF (2002) Z Anorg Allg Chem 628:2537–2541
111. Kraus F, Hanauer T, Korber N (2005) Angew Chem Int Ed 44:7200–7204
112. Korber N, Richter F (1997) Angew Chem Int Ed Engl 36:1512–1514
113. Reil M, Korber N (2007) Z Anorg Allg Chem 633:1599–1602
114. Pfisterer K (1999) Strukturchemie ammoniakreicher Solvatkristalle: Produkte der Reduktion von ungesättigten Kohlenwasserstoffen und Pnicogenchalkogeniden mit Alkalimetallen und Rekristallisation von Polyphosphiden in flüssigem Ammoniak Diploma thesis Universität Regensburg, Regensburg
115. Korber N, Daniels J (1996) Dalton Trans:1653–1658
116. Korber N, von Schnering H-G (1996) Chem Ber 129:155–159
117. Korber N, Daniels J (1996) Acta Crystallogr C52:2454–2457
118. Korber N, Daniels J (1996) Helv Chim Acta 79:2083–2087
119. Korber N, Daniels J (1999) Z Anorg Allg Chem 625:189–191
120. Hanauer T, Grothe M, Reil M, Korber N (2005) Helv Chim Acta 88:950–961
121. Korber N, von Schnering H-G (1997) Z Kristallogr NCS 212:85–86
122. Driess M, Merz K, Pritzkow H, Janoschek R (1996) Angew Chem Int Ed Engl 35:2507–2510
123. Hubler K, Becker G (1998) Z Anorg Allg Chem 624:483–496
124. Hanauer T, Aschenbrenner JC, Korber N (2006) Inorg Chem 45:6723–6727
125. Miluykov V, Kataev A, Sinyashin O, Lönnecke P, Hey-Hawkins E (2006) Z Anorg Allg Chem 632:1728–1732
126. Korber N, Daniels J (1996) Polyhedron 15:2681–2688
127. Korber N, Daniels J, von Schnering H-G (1996) Angew Chem Int Ed Engl 35:1107–1110
128. Korber N, Richter F (1996) Chem Commun:2023–2024
129. Korber N, Daniels J (1996) Z Anorg Allg Chem 622:1833–1838
130. Knettel D, Reil M, Korber N (2001) Z Naturforsch 56b:965–969
131. Belin CHE (1980) J Am Chem Soc 102:6036–6040
132. Hanauer T, Korber N (2006) Z Anorg Allg Chem 632:1135–1140
133. Bolle U, Tremel W (1992) J Chem Soc Chem Commun:91–93
134. Korber N (1997) Phosphorus Sulfur Silicon Relat Elem 125:339–346
135. Haushalter RC, Eichhorn BW, Rheingold AL, Geib S (1988) Chem Commun:1027–1028
136. Fritz G, Schneider HW, Hönle W, von Schnering H-G (1988) Z Anorg Allg Chem 43:561–566
137. Tebbe KF, Feher M, Baudler M (1985) Z Kristallogr 170:180–181
138. Rios D, Sevov SC (2010) Inorg Chem 49:6396–6398
139. Wade K (1972) Nucl Chem Lett 8:559
140. Rosdahl J, Fässler TF, Kloo L (2005) Eur J Inorg Chem:2888–2894
141. Wang J-Q, Wahl B, Fässler TF (2010) Angew Chem Int Ed 49:6592–6595
142. Richards AF, Hope H, Power PP (2003) Angew Chem Int Ed 42:4071–4074
143. Kirchner P, Huttner G, Heinze K, Renner G (1998) Angew Chem Int Ed 37:1664–1666
144. Renner G, Kircher P, Huttner G, Rutsch P, Heinze K (2001) Eur J Inorg Chem:973–980
145. Kesanli B, Fettinger J, Eichhorn BW (2001) Angew Chem Int Ed 43:6242–6247
146. Rivard E, Steiner J, Fettinger JC, Giuliani JR, Augustine MP, Power PP (2007) Chem Commun:4919–4921
147. Prabusankar G, Kempter A, Gemel C, Schröter M-K, Fischer RA (2008) Angew Chem Int Ed 47:7234–7237
148. Eichler BE, Power PP (2001) Angew Chem Int Ed 40:796–797

149. Wiberg N, Lerner H-W, Wagner S, Nöth H (1999) Z Naturforsch 54b:877–880
150. Kocak FS, Zavalij PY, Lam YF, Eichhorn BW (2009) Chem Commun:4197–4199
151. Chapman DJ, Sevov SC (2008) Inorg Chem 47:6009–6013
152. Gillett-Kunnath MM, Petrov I, Sevov SC (2010) Inorg Chem 49:721–729
153. Richards AF, Eichler BE, Brynda M, Olmstead MM, Power PP (2005) Angew Chem Int Ed 44:2546–2549
154. Goicoechea JM, Sevov SC (2006) Organomet 25:4530–4536
155. Zhou B, Denning MS, Jones C, Goicoechea JM (2009) Dalton Trans:1571–1578
156. Zhou B, Denning MS, Chapman TAD, Goicoechea JM (2009) J Am Chem Soc 48:2899–2907
157. Downing DO, Zavalij P, Eichhorn BW (2010) Eur J Inorg Chem:890–894
158. Eichhorn BW, Haushalter RC (1988) J Am Chem Soc 110:8704–8706
159. Kesanli B, Fettinger J, Eichhorn BW (2001) Chem Eur J 7:5277–5285
160. Campbell J, Mercier HPA, Franke H, Santry DP, Dixon DA, Schrobilgen GJ (2002) Inorg Chem 41:86–107
161. Yong L, Hoffmann SD, Fässler TF (2005) Z Kristallogr NCS 220:53–57
162. Schrenk C, Schellenberg I, Pöttgen R, Schnepf A (2010) Dalton Trans:1872–1876
163. Scharfe S, Fässler TF, Stegmaier S, Hoffmann SD, Ruhland K (2008) Chem Eur J 14:4479–4483
164. Scharfe S (2010) Reaktionen mit Zintl Ionen. Dissertation Thesis, Technische Universiät München, München
165. Kesanli B, Fettinger J, Gardner DR, Eichhorn BW (2002) J Am Chem Soc 124:4779–4786
166. Brynda M, Herber R, Hitchcock PB, Lappert MF, Nowik I et al (2006) Angew Chem Int Ed 45:4333–4337
167. Esenturk EN, Fettinger JC, Eichhorn BW (2006) J Am Chem Soc 128:12–13
168. Kesanli B, Halsig JE, Zavalij P, Fettinger JC, Lam YF, Eichhorn BW (2007) J Am Chem Soc 129:4567–4574
169. Sun ZM, Xiao H, Li J, Wang LS (2007) J Am Chem Soc 129:9560–9561
170. Kocak FS, Zavalij P, Lam Y-F, Eichhorn BW (2008) Inorg Chem 47:3515–3520
171. Wiberg N, Hochmuth W, Nöth H, Appel A, Schmidt-Amelunxen M (1996) Angew Chem Int Ed 35:1333–1334
172. Richards AF, Brynda M, Olmstead MM, Power PP (2004) Organomet 23:2841–2844
173. Schnepf A, Köppe R (2003) Angew Chem Int Ed 42:911–913
174. Schnepf A, Drost C (2005) Dalton Trans:3277–3280
175. Benda C (2009) Vinylierungsreaktionen am nonagermanium-polyanion. Diploma Thesis, Technische Universität München, München
176. Ugrinov A, Sevov SC (2004) Chem Eur J 10:3727–3733
177. Hull MW, Sevov SC (2009) J Am Chem Soc 131:9026–9037
178. Hull MW, Sevov SC (2007) Inorg Chem 46:10953–10955
179. Hull MW, Sevov SC (2007) Angew Chem Int Ed 46:6695–6698
180. Ugrinov A, Sevov SC (2002) J Am Chem Soc 124:2442–2443
181. Ugrinov A, Sevov SC (2003) J Am Chem Soc 125:14059–14064
182. Hull MW, Ugrinov A, Petrov I, Sevov SC (2007) Inorg Chem 46:2704–2708
183. Schnepf A (2003) Angew Chem Int Ed 42:2624–2625
184. Schenk C, Schnepf A (2009) Chem Commun:3208–3210
185. Scharfe S, Fässler TF (2010) Eur J Inorg Chem:1207
186. Goicoechea JM, Sevov SC (2006) J Am Chem Soc 128:4155–4161
187. Sun Z-M, Zhao Y-F, Li J, Wang L-S (2009) J Clust Sci 20:601–609
188. Spiekermann A, Hoffmann SD, Kraus F, Fässler TF (2007) Angew Chem Int Ed 46:1638–1640
189. Schenk C, Henke F, Santiso-Quinones G, Krossing I, Schnepf A (2008) Dalton Trans:4436–4441
190. Schenk C, Schnepf A (2007) Angew Chem Int Ed 46:5314–5316

191. Henke F, Schenk C, Schnepf A (2009) Dalton Trans:9141–9145
192. Nienhaus A, Hauptmann R, Fässler TF (2002) Angew Chem Int Ed 41:3213–3215
193. Boeddinghaus MB, Hoffmann SD, Fässler TF (2007) Z Anorg Allg Chem 633:2338–2341
194. Denning MS, Goicoechea JM (2008) Dalton Trans:5882–5885
195. Schnepf A, Schenk C (2006) Angew Chem Int Ed 45:5373–5376
196. Sekiguchi A, Ishida Y, Kabe Y, Ichinohe M (2002) J Am Chem Soc 124:8776–8777
197. Schnepf A (2007) Chem Commun:192–194
198. Schnepf A, Schenk C (2008) Chem Commun:4643–4645
199. Spiekermann A, Hoffmann SD, Fässler TF, Krossing I, Preiss U (2007) Angew Chem Int Ed 46:5310–5313
200. Goicoechea JM, Sevov SC (2005) Angew Chem Int Ed 44:4026–4028
201. Goicoechea JM, Sevov SC (2005) J Am Chem Soc 127:7676–7677
202. Wang J-Q, Stegmaier S, Fässler Thomas F (2009) Angew Chem Int Ed 48:1998
203. Zhou B, Denning MS, Kays DL, Goicoechea JM (2009) J Am Chem Soc 131:2802–2803
204. Stegmaier S (2007) Theoretische und experimentelle Untersuchungen zu Ge9-Polyanionen. Diploma Thesis, Technische Universität München, München
205. Hershaft A, Corbett JD (1963) Inorg Chem 2:979
206. von Benda H, Simon A, Bauhofer W (1978) Z Anorg Allg Chem:53
207. Krebs B, Hucke M, Brendel C (1982) Angew Chem Int Ed Engl 21:445–446
208. Corbett JD (1968) Inorg Chem 17:198
209. Beck J, Brendel C, Bengtsson-Kloo L, Krebs B, Mummert M et al (1996) Chem Ber 129:1219
210. Baudler M, Akpapoglou S, Ouzounis D, Wasgestian F, Meinigke B et al (1988) Angew Chem Int Ed Engl 27:280–281
211. Ichinohe M, Toyoshima M, Kinjo R, Sekiguchi A (2003) J Am Chem Soc 125:13328–13329
212. Fischer G, Huch V, Mayer P, Vasisht SK, Veith M, Wiberg N (2005) Angew Chem Int Ed 44:7884–7887
213. Joseph S, Hamberger M, Mutzbauer F, Härtl O, Meier M, Korber N (2009) Angew Chem Int Ed 48:8770–8772
214. Yong L, Hoffmann SD, Fässler TF, Riedel S, Kaupp M (2005) Angew Chem Int Ed 44:2092–2096
215. Eichhorn BW, Haushalter RC (1990) Chem Commun:937–939
216. Yong L, Hoffmann SD, Fässler TF (2005) Eur J Inorg Chem:3663–3669
217. Zhou B, Denning MS, Chapman TAD, McGrady JE, Goicoechea JM (2009) Chem Commun:7221–7223
218. Esenturk EN, Fettinger J, Eichhorn BW (2005) Chem Commun:247–249
219. Esenturk EN, Fettinger J, Eichhorn BW (2006) J Am Chem Soc 128:9178–9186
220. Esenturk EN, Fettinger J, Lam Y-F, Eichhorn BW (2004) Angew Chem Int Ed 43:2132–2134
221. von Schnering H-G, Wolf J, Weber D, Ramirez R, Meyer T (1986) Angew Chem Int Ed Engl 25:353–353
222. Foust AS, Foster MS, Dahl LF (1969) J Am Chem Soc 91:5631–5633
223. Rheingold AL, Foley MJ, Sullivan PJ (1982) J Am Chem Soc 104:4727–4729
224. Mattamana SP, Promprai K, Fettinger JC, Eichhorn BW (1998) Inorg Chem 37:6222–6228
225. Charles S, Eichhorn BW, Rheingold AL, Bott SG (1994) J Am Chem Soc 116:8077–8086
226. Eichhorn BW, Haushalter RC, Huffman JC (1989) Angew Chem Int Ed Engl 28:1032–1033
227. Kesanli B, Charles S, Lam YF, Bott SG, Fettinger JC, Eichhorn BW (2000) J Am Chem Soc 122:11101–11107
228. Knapp C, Zhou B, Denning MS, Rees NH, Goicoechea JM (2010) Dalton Trans 39:426–436
229. Moses MJ, Fettinger J, Eichhorn BW (2002) J Am Chem Soc 124:5944–5945
230. Eichhorn BW, Mattamana SP, Gardner DR, Fettinger JC (1998) J Am Chem Soc 120:9708–9709

231. Ahlrichs R, Fenske D, Fromm K, Krautscheid H, Krautscheid U, Treutler O (1996) Chem Eur J 2:238–244
232. Scherer OJ (2000) Angew Chem Int Ed 39:1029–1030
233. Santiso-Quinones G, Reisinger A, Slattery J, Krossing I (2007) Chem Commun:5046
234. Peruzzini M, Gonsalvi L, Romerosa A (2005) Chem Soc Rev 34:1038–1047
235. Groer T, Baum G, Scheer M (1998) Organometallics 17:5916–5919
236. Xiong Y, Yao S, Brym M, Driess M (2007) Angew Chem Int Ed 46:4511–4513
237. Fox AR, Wright RJ, Rivard E, Power PP (2005) Angew Chem Int Ed 44:7729–7733
238. Weigand JJ, Holthausen M, Fröhlich R (2009) Angew Chem Int Ed 48:295–298
239. Scherer OJ, Vondung J, Wolmershauser G (1989) Angew Chem Int Ed Engl 28:1355–1357
240. Scheer M, Herrmann E, Sieler J, Oehme M (1991) Angew Chem Int Ed Engl 30:969–971
241. Krossing I (2001) J Am Chem Soc 123:4603–4604
242. Urnezius E, Brennessel WW, Cramer CJ, Ellis JE, von Ragué Schleyer P (2002) Science 295:832–834
243. Scherer OJ, Brück T (1987) Angew Chem Int Ed Engl 26:59
244. Scherer OJ, Sitzmann H, Wolmershauser G (1985) Angew Chem Int Ed Engl 24:351–353
245. Dai FR, Xu L (2006) Inorg Chim Acta 359:4265–4273
246. Aschenbrenner JC, Korber N (2004) Z Anorg Allg Chem 630:31–32
247. Charles S, Fettinger JC, Bott SG, Eichhorn BW (1996) J Am Chem Soc 118:4713–4714
248. Charles S, Fettinger JC, Eichhorn BW (1996) Inorg Chem 35:1540–1548
249. Cossairt BM, Cummins CC (2008) Angew Chem Int Ed 47:169–172
250. Korber N, von Schnering H-G (1995) J Chem Soc Chem Commun:1713–1714
251. Chan WTK, Garcia F, Hopkins AD, Martin LC, McPartlin M, Wright DS (2007) Angew Chem Int Ed 46:3084–3086
252. Fritz G, Hoppe KD, Hönle W, Weber D, Mujica C et al (1983) J Organomet Chem 249:63–80
253. Fritz G, Layher E, Hönle W, von Schnering H-G (1991) Z Anorg Allg Chem 595:67–94
254. Konchenko SN, Pushkarevsky NA, Gamer MT, Köppe R, Schnöckel H, Roesky PW (2009) J Am Chem Soc 131:5740–5741
255. Scherer OJ, Berg G, Wolmershauser G (1996) Chem Ber 129:53–58
256. Barr ME, Adams BR, Weller RR, Dahl LF (1991) J Am Chem Soc 113:3052–3060
257. Scheer M, Becker U, Matern E (1996) Chem Ber 129:721–724
258. Ye YZ, Xu L (2008) Chin J Struc Chem 27:75–77
259. Korber N, Daniels J (1997) Inorg Chem 36:4906–4908
260. Masuda JD, Schoeller WW, Donnadieu B, Bertrand G (2007) J Am Chem Soc 129:14180–14181
261. Baudler M (1982) Angew Chem Int Ed Engl 21:492–512
262. Goicoechea JM, Hull MW, Sevov SC (2007) J Am Chem Soc 129:7885–7893
263. Breunig HJ, Rösler R, Lork E (1997) Angew Chem Int Ed Engl 36:2819–2821
264. Stark JL, Harms B, Guzman-Jimenez I, Whitmire KH, Gautier R et al (1999) J Am Chem Soc 121:4409–4418
265. Breunig HJ, Burford N, Rösler R (2000) Angew Chem Int Ed 39:4148–4150
266. Bolle U, Tremel W (1994) J Chem Soc Chem Commun:217–219
267. Charles S, Eichhorn BW, Bott SG (1993) J Am Chem Soc 115:5837–5838
268. Moses MJ, Fettinger JC, Eichhorn BW (2007) Inorg Chem 46:1036–1038
269. Xu L, Ugrinov A, Sevov SC (2001) J Am Chem Soc 123:4091–4092
270. Whitmire KH, Churchill MR, Fettinger JC (1985) J Am Chem Soc 107:1056–1057
271. Groer T, Scheer M (2000) Organometallics 19:3683–3691
272. Goicoechea JM, Sevov SC (2006) Angew Chem Int Ed 45:5147–5150
273. Lips F, Dehnen S (2009) Angew Chem Int Ed 48:6435–6438
274. Aoyagi S, Nishibori E, Sawa H, Sugimoto K, Takata M et al (2010) Nat Chem 2:678–683

Struct Bond (2011) 140: 133–154
DOI: 10.1007/430_2010_22

Published online: 27 October 2010

Germanium-Based Porous Semiconductors from Molecular Zintl Anions

Gerasimos S. Armatas and Mercouri G. Kanatzidis

Abstract This review highlights how molecular Zintl compounds can be used to create new materials with a variety of novel opto-electronic and gas absorption properties. The generality of the synthetic approach described in this chapter on coupling various group-IV Zintl clusters provides an important tool for the design of new kinds of periodically ordered mesoporous semiconductors with tunable chemical and physical properties. We illustrate the potential of Zintl compounds to produce highly porous non-oxidic semiconductors, and we also cover the recent advances in the development of mesoporous elemental-based, metal-chalcogenide, and binary intermetallic alloy materials. The principles behind this approach and some perspectives for application of the derived materials are discussed.

Keywords Chalcogenide · Germanium compound · Mesoporous semiconductor · Self-assembly · Zintl compound

Contents

G.S. Armatas
Department of Materials Science and Technology, University of Crete, Heraklion 71003, Greece
e-mail: garmatas@materials.uoc.gr

M.G. Kanatzidis (✉)
Department of Chemistry, Northwestern University, Evanston, IL 60208, USA
e-mail: m-kanatzidis@northwestern.edu

1 Introduction

Porous semiconductors represent an emerging class of functional materials with fascinating properties and application potential in photocatalysis, separations, sensing, and optoelectronics. The materials offer co-existence of electronic and optical characteristics, with porosity as a unique combination which is difficult to accomplish in discrete nanoparticles or conventional nanoporous materials. Porous materials possessing a semiconducting framework structure with wall thickness and pore sizes in the nanoscale dimension (1–100 nm) can be viewed as inorganic replicas of a regular array of spatially arranged quantum dots [1, 2]. Such structures can exhibit electronic and optical characteristics similar to those of discrete nanodots, e.g., size-dependent quantum confinement [3–6]. Furthermore, the existence of regular porosity can induce useful functionalities, including size-selective photocatalysis, membrane separation, and photochemical or electrochemical sensing.

In contrast to mesoporous oxides, mesoporous non-oxidic frameworks are challenging to prepare because of the difficulty in growing or organizing appropriate precursors in ordered forms such as mesophases and thin films. Fabrication techniques previously developed for non-oxidic semiconductive composites are based on "soft" supramolecular templating routes or "hard" nanocasting approaches [7, 8]. The liquid-crystal templating has a suitable technique for achieving synthetic control on the nanoscale. This method involves electrostatic interactions and charge matching at the interface of the self-assembled surfactants and inorganic species. So far, this technique led to a wide variety of non-oxidic inorganic mesophases with remarkable pore architectures by controlling the composition of the reaction medium (i.e., surfactant chain length, Zintl cluster, solvent) [9].

The challenge of making porous semiconducting framework materials led to the development of synthetic routes involving the binding of Zintl clusters (i.e., dimeric $Sn_2Se_6^{4-}$ or adamantine-type $Ge_4S_{10}^{4-}$) with a variety of linking metal ions [9–16]. These include ordered mesostructured metal-chalcogenides such as C_nPyGaGeS, C_nPyInGeS, and C_nPyPtSnSe [13, 15, 16]. These materials possess a well-ordered mesostructured framework with wormhole, hexagonal, or cubic pore structure, and optoelectronic properties originate from the surfactant/chalcogenide composite. Therefore, they can be easily tuned by controlling the chemical composition and surfactant head group.

Microporous metal sulfides and selenides with open-framework structure have also been prepared [17–24]. These structures are crystalline and exhibit adsorption, ion-exchange, and sensing behavior that resemble those of small-pore Zeolites [25–27]. Recently, guest-free germanium ($\square_{24}Ge_{136}$, where $\square$ means an atom vacancy) and silicon ($\square_{24}Si_{136}$) crystalline clathrate-II-type structures have been reported [28, 29]. These structures were produced through the thermal decomposition of the Zintl compounds K_4Ge_9 and NaSi under vacuum or ambient conditions and remarkably are crystalline with an extended three-dimensional network of tetrahedrally coordinated Ge or Si atoms. Interestingly, the cages of these structures do not encapsulate guest atoms such as the other clathrates, but they are really empty. Although this is an important subclass of inorganic materials with open-framework structure, these solids are not porous (i.e., solids permeated by an interconnected and continuous network of pores or voids) and, therefore, they will not be covered further. This review highlights the potential of Zintl compounds in generating new and fascinating classes of porous materials. The concepts described here are general and apply to many building blocks. The review focuses particularly on the potential of Zintl anions composed of single elements such as Ge, Si, and Sn to serve as unique precursors to porous semiconductor materials. The idea of course can be extended to any molecular Zintl anion that can serve as a precursor.

2 Mesoporous Forms of Germanium

Germanium, like silicon, is an important semiconductor with considerable scientific interest because of its many practical technological applications [30–32]. As a result, a large number of synthetic procedures for Ge nanoparticles have been developed that affect size and shape of the final products. Although the synthesis of germanium nanoparticles is well established, the synthesis of Ge semiconductors with porous frameworks is not.

Nanostructured germanium structures with mesoporous channels running through them can represent a different form of framework materials with exciting physical properties. When bulk germanium is penetrated by a regular array of mesopore channels, nanosized walls can be created that can exhibit electronic and optical characteristics similar to those of discrete nanodots [33].

Recently, we and others demonstrated that appropriate germanide Zintl clusters in non-aqueous liquid-crystalline phases of cationic surfactants can assemble well-ordered mesostructured and mesoporous germanium-based semiconductors. These include mesostructured cubic gyroidal and hexagonal mesoporous Ge as well as ordered mesoporous binary intermetallic alloys and Ge-rich chalcogenide semiconductors.

The group-IV Zintl clusters represent an interesting class of molecules with fascinating electronic properties. They can easily be obtained in crystalline form by the solid-phase reduction of elements Si, Ge, Sn, or Pb with alkali metals. The high reactivity toward direct or metal-mediated oxidative coupling, flexible composition, and alterable solubility make these inorganic compounds attractive building block candidates for the fabrication of new materials. Some of the relevant building units include the deltahedral homoatomic nine-atom $E_9{}^{4-}$ (E = Si, Ge, Sn, Pb) anions, the functionalized $[EPh_2\text{-}Ge_9\text{-}EPh_2]^{2-}$ (E = Bi, Sb, Sn), and $[R_3E\text{-}(Ge_9)_2\text{-}ER_3]^{4-}$ (E = Ge, Sn; R = Me, Ph) clusters and the transition metal-substituted $[E_9M(CO)_3]^{4-}$ (E = Sn, Pb; M = Cr, Mo, W), $[Sn_9Pt_2(Ph_3)]^{2-}$, $[Ni_3@(Ge_9)_2]^{4-}$, $[Pt@Pb_{12}]^{2-}$, and $[Ge_8Zn]^{6-}$ clusters [34–39]. Typical structures of these anions are shown in Scheme 1.

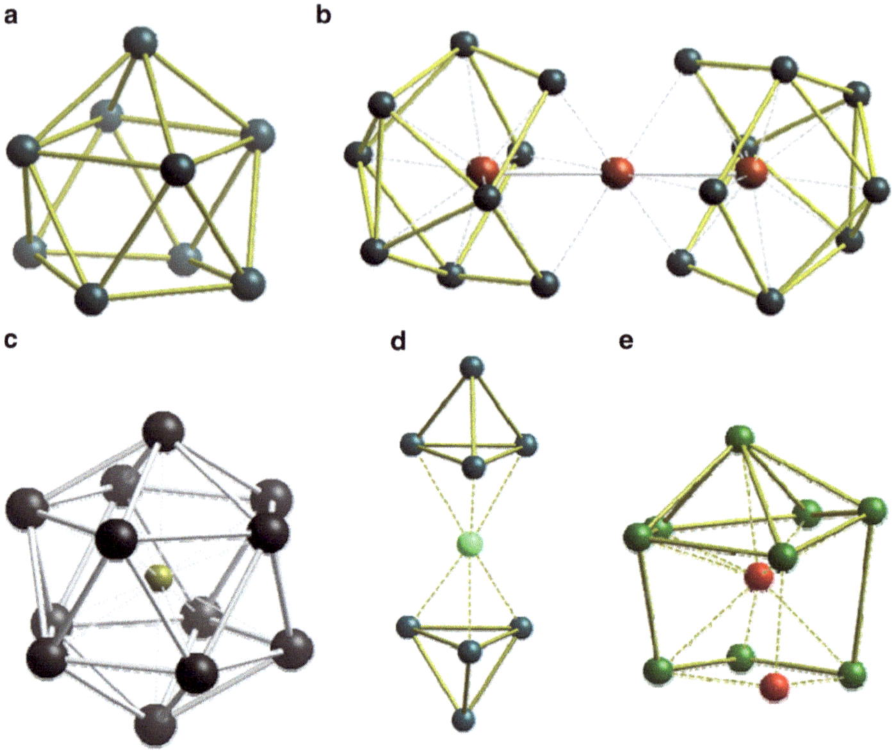

Scheme 1 Typical Zintl clusters (**a**) $[Ge_9]^{4-}$, (**b**) $[Ni_3@(Ge_9)_2]^{4-}$, (**c**) $[Pt@Pb_{12}]^{2-}$, (**d**) $[Ge_8Zn]^{6-}$ and (**e**) $[Sn_9Pt_2]^{2-}$ as building block candidates for the construction of porous semiconductors

2.1 Cubic Mesostructured Germanium

Covalently bonded Ge frameworks can be prepared from combining negatively charged germanium-based Zintl anions and positively charged Ge species or by oxidatively polymerizing anionic Zintl clusters. If these frameworks are built within an organized surfactant medium, ordered porous materials can be obtained. The proper precursors for this are $[Ge_9]^{4-}$ clusters and Ge(-IV) atoms [40]. The latter is derived from the simple Zintl compound Mg_2Ge and gives cubic framework when react with $GeCl_4$ in the presence of a structure-directing surfactant:

The chemistry of Scheme 2 produces a cubic pore structure with long-range periodicity and unit cell parameter (α_o) of 8.4 nm. The material show a relatively large number of Bragg peaks in the X-ray diffraction (XRD) pattern, which can be indexed as (211), (220), (321), (400), (420), (332), (422), (431), (611), and (543) Bragg diffraction peaks of the body-centered cubic $I\alpha$-$3d$ symmetry (Fig. 1).

Figure 2 presents typical TEM images obtained from MSU-Ge-1 and the corresponding fast-Fourier transforms (FFT) taken along a thin area of the images. These images clearly indicate a high ordered mesoporous structure in body-centered cubic

Scheme 2 Reaction steps for the assembly of cubic mesostructured MSU-Ge-1 (MSU: Michigan State University) germanium

$[C_{20}H_{41}\text{-}N(CH_2CH_2OH)_2CH_3]Br + Mg_2Ge$

↓ Formamide, N_2 atmosphere

"$[C_{20}H_{41}\text{-}N(CH_2CH_2OH)_2CH_3]_x\{Ge^{(-IV)}\}$" + $MgBr_2$

Organized micellar solution

↓ $GeCl_4$ linking agent

MSU-Ge-1

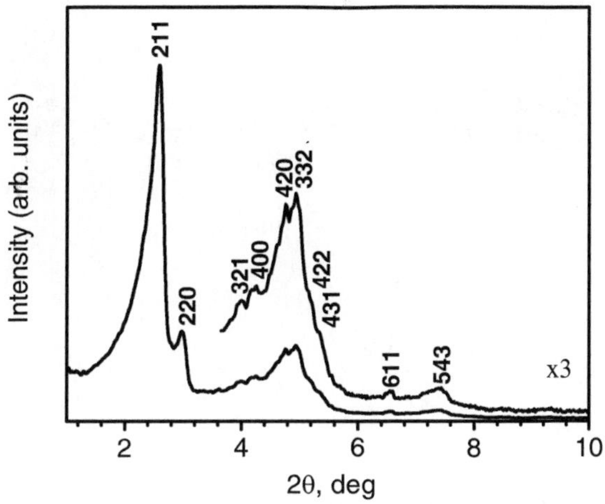

Fig. 1 Low-angle powder XRD pattern of cubic mesostructured germanium. The Bragg diffraction peaks consistent with the body-centered *Iα-3d* unit cell with $\alpha_o = 8.4$ nm

symmetry. From the TEM images, the size of the surfactant, the pore-to-pore distance from XRD, and the wall thickness of the Ge framework are estimated at ~1 nm. Despite the amorphous character of the Ge framework, pair distribution functional analysis [41] reveals a well-defined local structure of tetrahedral coordinated Ge atoms, similar to these of bulk polycrystalline and amorphous Ge. This material could not be made porous, because all attempts to remove the surfactant caused the framework to collapse, possibly because of its very thin walls (~1 nm).

A particularly interesting property of this mesostructured cubic Ge framework is the substantial blue-shift in optical absorption at 1.42 eV relative to 0.66 eV of bulk Ge. This large blue-shift can be understood by considering the change in the density of the electronic energy states caused by the substantial dimensional reduction of the Ge structure from the bulk Ge (infinity wall thickness) to an ~1 nm. Similar large blue-shifted band gaps by ~0.76 eV are also observed in Ge nanocrystals of ~4 nm in diameter, which is a consequence of quantum confinement effects [42]. Whether

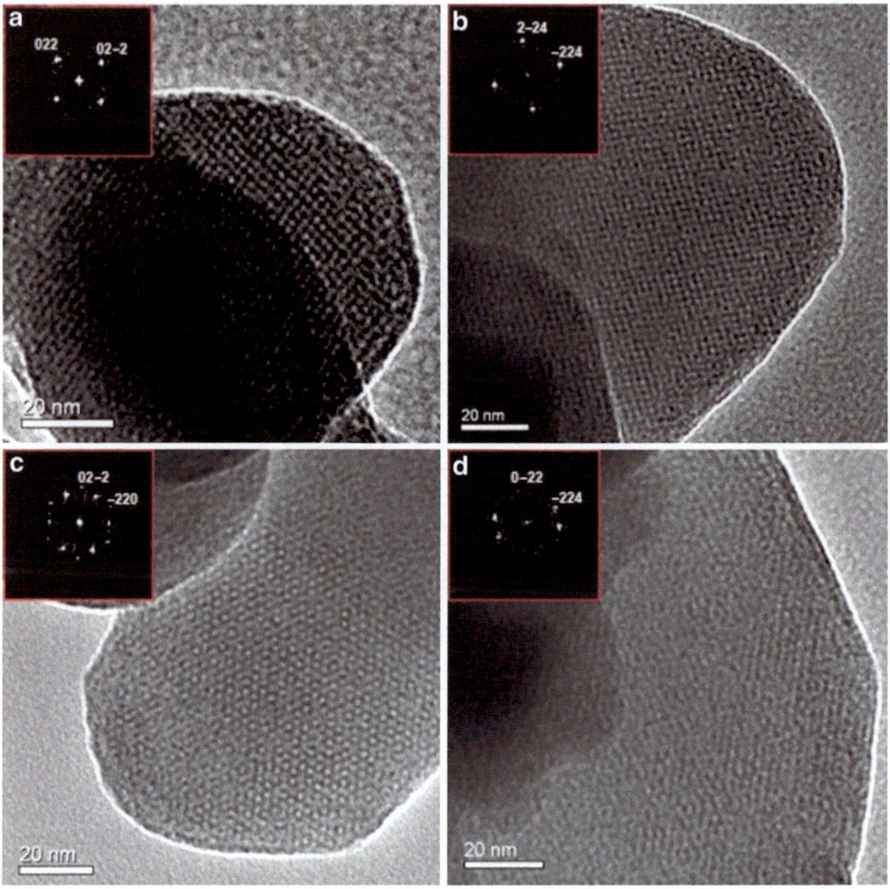

Fig. 2 Typical TEM images of mesostructured germanium MSU-Ge-1 semiconductor taken (**a**) along the [100] direction, (**b**) along the [110] direction, (**c**) along the [111] direction, and (**d**) along the [311] direction. The pore–pore spacings are consistent with those deduced from X-ray diffraction. Insets in each panel show FFT images taken along an area of the high-resolution TEM images. Reproduced with permission from [40]. Copyright 2006 Nature Publishing Group

there is a quantum confinement in the porous amorphous Ge networks described in this chapter is an open question and needs to be addressed with proper physicochemical experiments. This question is further complicated by the fact that very little is known about quantum confinement in nanoparticles of amorphous compounds.

2.2 *Hexagonal Mesoporous Germanium*

Using a different Zintl anion as building block, we and others prepared mesoporous germanium with well-ordered pore structure and high internal surface area [43, 44].

A new family of hexagonal mesoporous all-germanium semiconductors was prepared by the surfactant-assisted cross-linking polymerization reaction of Zintl $[Ge_9]^{4-}$ anions (Scheme 1) with Ge(IV) bridges in formamide/ethylenediamine solution (1) [43].

$$\begin{aligned} &\text{R-N}(CH_3)_2(CH_2CH_2OH)\text{Br}+K_4Ge_9+GeCl_4 \rightarrow \\ &\qquad\qquad \{\text{R-N}(CH_3)_2(CH_2CH_2OH)\}_{0.8}Ge_{10}+KCl+KBr \\ &\qquad\qquad\qquad\qquad \text{MSU-Ge-2} \\ &(R = C_nH_{2n+1};\, n = 18, 20, 22) \end{aligned} \tag{1}$$

Unlike the cubic materials above, this material can be rendered porous after carefully removing the surfactants inside the pores via a two-step ion-exchange/thermal treatment procedure. Namely, the large surfactants can be replaced with small NH_4^+ cations without framework collapse, which is subsequently removed in the volatile form of NH_3 by mild heating to give hexagonal mesoporous structure. That the framework does not shrink around the small ammonium ions attests to its rigidity. This is to be contrasted with the reversible contraction/expansion of the chalcogenide framework of cubic Pt/Sn_2Se_6 upon ion exchange with small and large cations, respectively [45]. The chalcogenide framework is flexible presumably because of the two-bonded chalcogen bridges Pt-Se-Sn that can act as hinges, allowing conformational changes. On the other hand, the all-Ge network contains multiply bonded atoms which increase its structural rigidity [46].

XRD and TEM analysis on template-removed MSU-Ge-2 evidenced the presence of a well-defined, long-range periodicity of the hexagonal pore structure (Fig. 3). The low-angle powder XRD pattern of "as-prepared" and "template-removed" mesoporous MSU-Ge-2 indicates a pore periodicity of ~4.8 and ~4.0 nm, respectively. The pore-to-pore distance (4.0 nm) determined from XRD

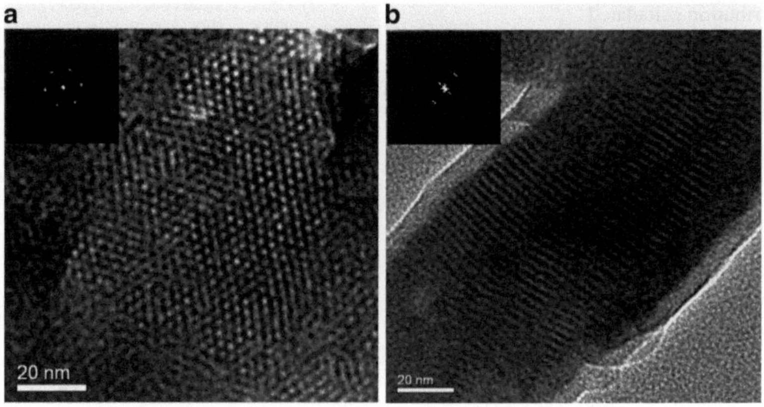

Fig. 3 TEM and (*insets*) FFT images of mesoporous MSU-Ge-2 obtained by heat-treating the NH_4^+-exchanged material. The images shows (**a**) a periodic hexagonal array along the [001] direction and (**b**) a periodic array of parallel pore channels along the [110] direction. Reproduced with permission from [43]. Copyright 2007 American Association for the Advancement of Science

together with the pore diameter of ~2.1 nm determined from nitrogen physisorption measurements gave estimation from the mesoporous framework wall thickness of ~1.9 nm. The surface of the porous Ge framework was found to be passivated with Ge-H terminal bonds according to FT-IR spectroscopy.

The N_2 adsorption–desorption isotherms for MSU-Ge-2 show a type-IV adsorption branch associated with a well-defined capillary condensation step at P/Po ~0.13, characteristic of uniform mesopores (Fig. 4). The adsorption data indicate a very high Brunauer–Emmett–Teller (BET) surface area of 363 m^2/g and a pore volume of 0.23 cm^3/g. Given that the Ge mesostructure is much heavier than the corresponding silica, this surface area is actually equivalent to silica with a surface area of 1,316 m^2/g.

The pore size in hexagonal MSU-Ge-2 can be easily controlled by adjusting the length of the surfactant chain. For example, surfactants with different chain length $C_{18}H_{37}$- and $C_{22}H_{45}$- gave hexagonal mesostructures with surface areas of 109 and 364 m^2/g, pore volume of 0.09 and 0.40 cm^3/g, and uniform pore size of ~1.7 and ~2.2 nm, respectively.

The porous Ge mesostructures exhibit energy band gaps that are much wider than that of bulk crystalline or amorphous Ge. In Fig. 5, the optical absorption

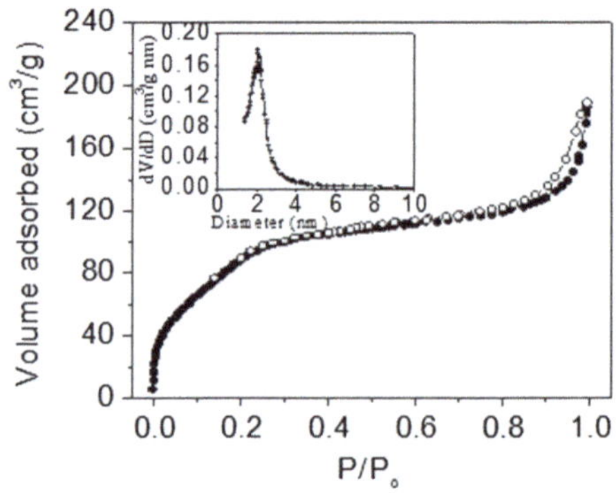

Fig. 4 Nitrogen adsorption–desorption isotherms at 77 K of heat-treated NH_4^+-exchanged MSU-Ge-2 (*solid circles*, adsorption data; *open circles*, desorption data). The hysteresis observed at P/Po > 0.8 is due to the voids between the agglomerated particles. (*Inset*) BJH pore size distribution calculated from the adsorption branch of the isotherm

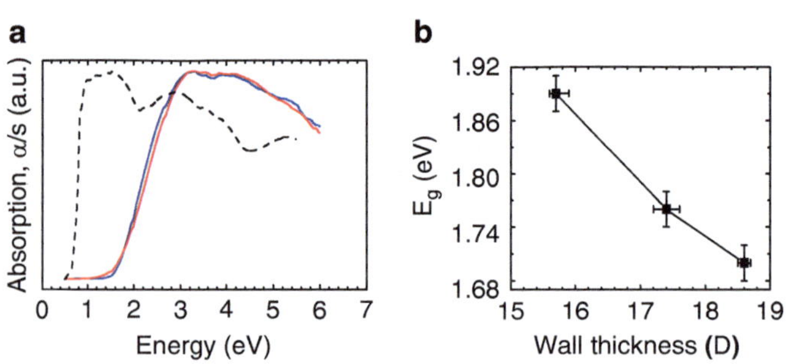

Fig. 5 (**a**) Optical absorption spectra of as-prepared (*blue*) and mesoporous MSU-Ge-2 (*red*) and bulk polycrystalline Ge (*dashed line*). (**b**) Size-dependence of the energy band gap of MSU-Ge-2 as a function of framework wall thickness

spectra of mesoporous MSU-Ge-2 semiconductor show energy gap transition at 1.70 eV, which is consequently blue-shifted compared to bulk Ge (0.66 eV). This enormous blue-shift observed in the porous material is attributed to the considerable dimensional reduction of the wall thickness (~1.9 nm) in MSU-Ge-2 compared to the bulk phase. Interestingly, adjusting the reaction conditions of the MSU-Ge-2, the wall thickness of the resulting Ge-framework can be tuned, and this is reflected in the systematic blue-shift of the energy band gap (Fig. 5b). The size-dependence of the energy band gap observed in mesoporous MSU-Ge-2 mirrors the quantum-confinement-induced blue-shift in the energy gap observed in Ge nanocrystals. In fact, this trend is analogous to the well-known ability to tune the average size of nanocrystals by adjusting the synthetic conditions to obtain size-dependent physical behavior [6, 47].

Upon using a different reaction chemistry, the synthesis of another type of porous Ge has been reported by Tolbert and co-workers [44]. They used oxidative coupling of linear polymer chain of ${}^{1}_{\infty}\{Ge_9^{2-}\}$ derived from K_2Ge_9 compound with ferrocenium cation to produce nanoporous hexagonal germanium. This reaction was carried out in ethylenediamine solution in the presence of structure-directing amphiphilic surfactants. The obtained material features a porous structure with slit-like pores and large internal surface of 500 m^2/g and an energy gap transition at ~1.4 eV. It was purported that the framework structure is made entirely of tetrahedrally coordinated Ge atoms.

The mesoporous forms of germanium that derive from the above chemistry are very air sensitive and rapidly convert to germanium suboxides GeO_x upon exposure in air for a short time (<1 min). This is expected since almost all Ge atoms of the framework lie at or near the surface and the Ge–Ge bond is susceptible in oxidation. The formation of GeO_x involves the conversion of Ge–Ge bonds to Ge–O–Ge moieties and seems to be a homogeneous process. This causes a systematic blue-shift of the energy band gap, possibly due to the size-confinement effect.

High surface area hexagonal mesoporous Ge also can be prepared with oxidative self-polymerization chemistry of $[Ge_9]^{4-}$ clusters [48]. This synthetic route does not require external oxidants such as ferrocenium or linking Ge(IV) centers and occurs in the presence of cationic surfactant (*N*-eicosane-*N*,*N*-dimethyl-*N*-(2-hydroxyethyl)ammonium bromide, EDMHEABr) as structure-directing agent. The polymerization reaction proceeds through the slow oxidative coupling of (Ge_9)-clusters and seems to be accompanied by a two-electron process (2). The electron acceptors in this case appear to be the surfactant molecules or the solvent.

$$\begin{aligned} &\mathrm{EDMHEABr} + \mathrm{K_4Ge_9} + \mathrm{xNH_2(CH_2)_2NH_2} \rightarrow \mathrm{(EDMHEA)[Ge_9]} + \\ &\mathrm{xNH_2(CH_2)_2NH^-} + \mathrm{(x/2)H_2} + \mathrm{KBr} \end{aligned} \tag{2}$$

Small-angle X-ray scattering (SAXS) diffraction and TEM images highlight the mesoporous nature of NU-Ge-1, revealing a well-ordered arrangement of pores in hexagonal *p6mm* symmetry (Fig. 6). As in the previous mesoporous version of Ge, here the wall structure of NU-Ge-1 does not possess long-range atomic order. This is typical for surfactant-templated mesostructures in general. X-ray diffuse scattering and pair distribution function analysis indicate that the local structure of the inorganic framework is well defined with interconnected –(Ge_9)– deltahedral clusters.

Nitrogen physisorption measurements indicate large internal surface area of 402 m^2/g and quite narrow distribution of pore sizes with peak maximum at ~3.1 nm (Fig. 7). The framework wall thickness was estimated to be ~1.9 nm. The oxidation state of Ge framework as probed with X-ray photoelectron spectroscopy (XPS) and time-of-flight secondary ion mass spectroscopy (ToF-SIMS) techniques is close to zero.

The NU-Ge-1 exhibits optical absorption effects that are tunable by changing the wall thickness of the germanium framework. The pore structure of this material can be both active and sensitive toward electron-transfer adsorbates. We have observed that the electronic structure of the NU-Ge-1 can be altered in a controllable manner

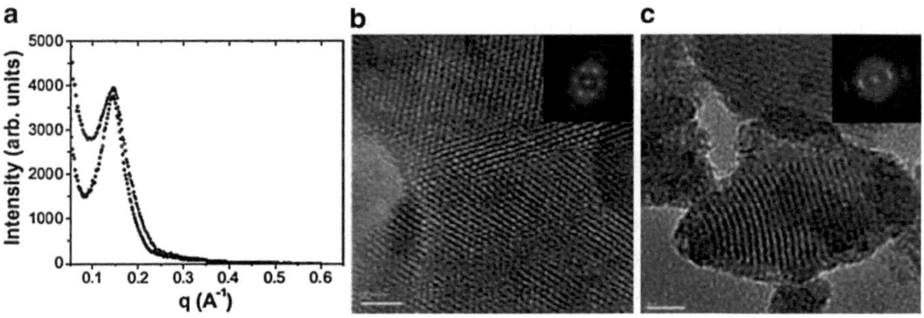

Fig. 6 (**a**) SAXS diffraction patterns for as-prepared containing surfactant (*closed symbols*) and mesoporous NU-Ge-1 (*open symbols*) materials, and (**b** and **c**) TEM images of NU-Ge-1 recorded along the [001] zone axis and along the [110] zone axis, respectively, showing the hexagonal array of uniform pore channels. *Insets*: the corresponding fast-Fourier transform (FFT) images. Reproduced with permission from [48]

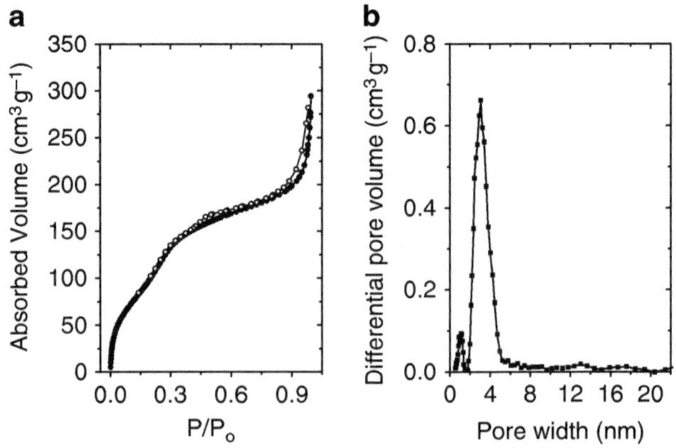

Fig. 7 (**a**) Nitrogen adsorption (*closed symbols*) and desorption (*open symbols*) isotherms at 77 K of mesoporous NU-Ge-1. (**b**) NLDFT pore size distribution calculated from the adsorption branch

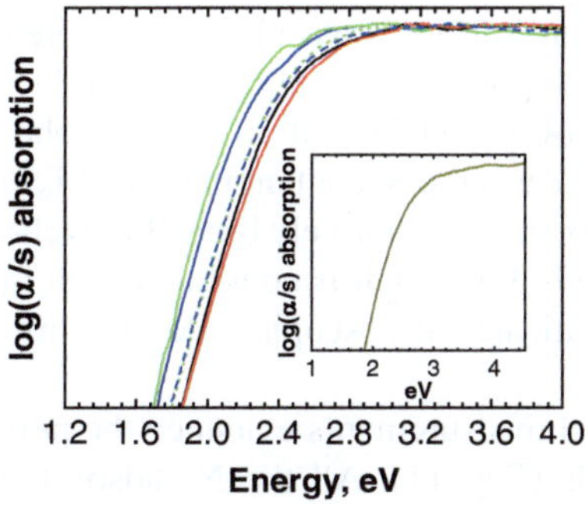

Fig. 8 Kubelka–Munk optical absorption spectra of as-prepared mesostructured (*black*) and mesoporous NU-Ge-1 (*red*) semiconductors and NU-Ge-1 incorporating into the pores TCNE (*blue*) and TTF (*green line*) organic molecules. The recovered optical adsorption spectra of NU-Ge-1 by encapsulation of TCNE-TTF complexes are also given (*dashed lines*). *Inset*: optical absorption spectrum of NU-Ge-1 encapsulating anthracene

when its mesopore surface interacts with organic molecules that engage in electron-transfer interactions (Fig. 8).

Namely, when electron-transfer adsorbates such as the electron-acceptor tetracyanoethylene (TCNE) and electron-donor tetrathiafulvalene (TTF) molecules interact with the inorganic framework, the energy gap of the mesoporous NU-Ge-1 (1.87 eV) is red-shifted to 1.71 and 1.64 eV, respectively. Indeed, this change in electronic structure is reversible and the optical adsorption onsets going to 1.83 eV upon formation of the "inactive" TTF-TCNE complex inside the pores. Incorporation of molecules without electron-acceptor or electron-donor properties such as anthracene did not affect the electronic structure of NU-Ge-1.

3 Mesoporous Ge/Si Alloy Semiconductors

The ability of using mixed $[Ge_{9\text{-}n}Si_n]^{4-}$ clusters as starting building blocks, which are soluble in ethylenediamine, allowed us to prepare mesoporous Ge/Si alloy semiconductors. These structures were synthesized as described above by the oxidative self-polymerization of mixed $[Ge_{9\text{-}n}Si_n]^{4-}$ clusters with the assistance of self-assembled cationic surfactants (3).

$$\begin{aligned}&\text{R-N}(CH_3)_2(CH_2CH_2OH)Br + K_4Ge_{9-n}Si_n + yNH_2(CH_2)_2NH_2 \rightarrow \\ &\quad [\text{R-N}(CH_3)_2(CH_2CH_2OH)]_xGe_{9-n}Si_n + KBr + yNH_2(CH_2)_2NH^- + (y/2)H_2\end{aligned} \tag{3}$$

Porosity can be created with ion-exchange using NH_4^+ that is subsequently removed as ammonia (NH_3) with gentle heating in vacuum. The materials possess a mesoporous framework that consists of interconnected Ge and Si atoms and they exhibit large and accessible surface areas (Table 1). Note that this is the first time where the synthesis of mesoporous Ge/Si alloy semiconductors is reported. Figure 9 shows the low-angle powder XRD patterns of template-removed mesoporous NU-GeSi-*N* (where *N* indicates the Si content in $Ge_{9-n}Si_n$ cluster; $N = 2n$) samples. All mesoporous structures show a relatively broad but well-defined diffraction peak at low scattering angles (1–3°), which is consistent with mesoscopic order. Direct observation with TEM indicates mesoscopic order in wormhole-like pore structure (Fig. 10).

Nitrogen physisorption measurements evidence the permanent mesoporosity of the NU-GeSi-*N* materials (Fig. 11). All the N_2 adsorption–desorption isotherms

Table 1 Analytical data and textural properties of mesoporous NU-GeSi-*N* materials

Sample	Atomic ratio[a] (Si:Ge)	Surface area (m^2g^{-1})	Pore volume[b] (cm^3g^{-1})	Pore size (nm)	d-spacing[c] (nm)	Energy gap (eV)
NU-GeSi-1	0.5:8.5	312	0.24	3.0	4.6	2.04
NU-GeSi-2	0.6:8.4	301	0.22	2.8	4.6	2.06
NU-GeSi-3	1.5:7.5	370	0.38	3.0	4.5	2.08
NU-GeSi-4	1.8:7.2	218	0.13	2.4	4.4	2.12

[a]XRF elemental data normalized to the nine-atom $Ge_{9-n}Si_n$ cluster
[b]Cumulative pore volume at $P/P_o = 0.95$
[c]d-spacing of mesoscopic order determined from the angular position of main diffraction peak

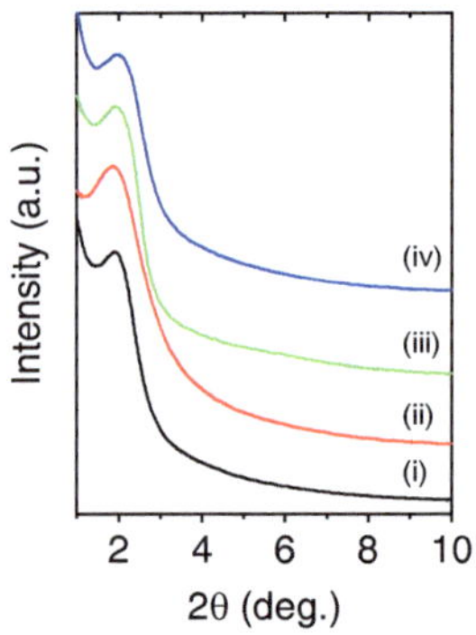

Fig. 9 Low-angle XRD patterns of template-free mesoporous materials: (i) NU-GeSi-1, (ii) NU-GeSi-2, (iii) NU-GeSi-3, and (iv) NU-GeSi-4

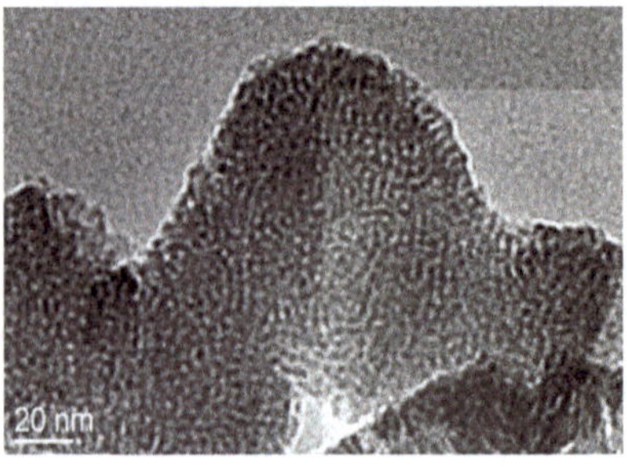

Fig. 10 Typical TEM image of mesoporous NU-GeSi-2 revealing a wormhole-type pore structure with a regular pore size

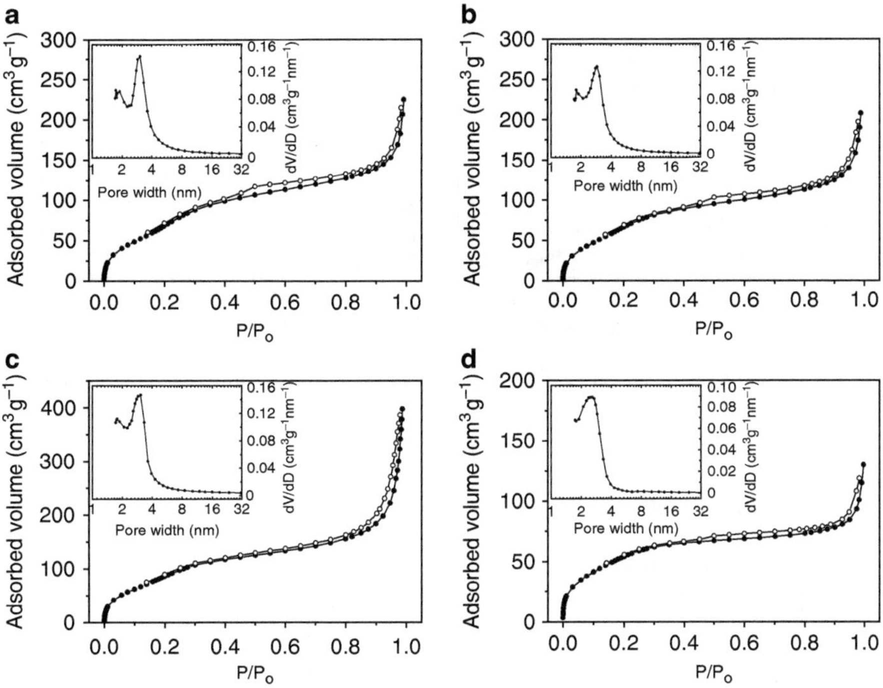

Fig. 11 Nitrogen adsorption–desorption isotherms of mesoporous (**a**) NU-GeSi-1, (**b**) NU-GeSi-2, (**c**) NU-GeSi-3, and (**d**) NU-GeSi-4 materials. *Inset*: BJH pore size distribution calculated from the adsorption branch

reveal typical type-IV adsorption branches with a well-defined "knee" at relative pressure P/P_o ~0.2 that is characteristic of well-ordered mesostructures. Analysis of the adsorption data with the BET method indicates surface areas in the range of 218–370 m^2/g and pore volumes in the range of 0.13–0.38 cm^3/g. The pore size distributions determined with the modified BJH method [49, 50] were found to be quite narrows with peak maxima in the range of 2.4–3.0 nm. All the analytical data and textural properties of the NU-GeSi-*N* materials are summarized in Table 1.

Figure 12 displays typical XPS spectra acquired on a mesoporous NU-GeSi-3 sample. The XPS analysis reveals a single Ge 3d core-level peak with binding energy at 29.1 eV, which is very close to that Ge 3d binding energy observed in zero-valent Ge atoms (29.0 eV) [51] and much lower than the Ge 3d binding energy of GeO_2 (32.5 eV) [52]. Similarly, the XPS spectrum of silicon has Si $2p_{3/2}$ core-level binding energy at 100.6 eV, implying that the silicon atoms are in a low oxidation state [53].

These materials feature a well-defined electronic structure that is consistent with (but does not prove) quantum confinement derived from the very thin wall thickness (Fig. 13). The optical absorption spectra of NU-GeSi-1 show sharp absorption onsets in the energy range from ~2.04 to ~2.12 eV. Crystalline and amorphous bulk Si have a band gap of 1.1 and 1.5 eV, respectively [54]. The energy band gap

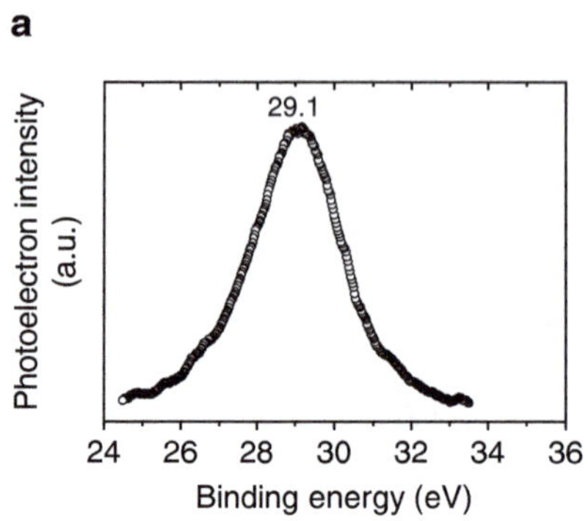

Fig. 12 Typical high-resolution (**a**) Ge 3d and (**b**) Si $2p_{3/2}$ photoelectron spectra of mesoporous NU-GeSi-3

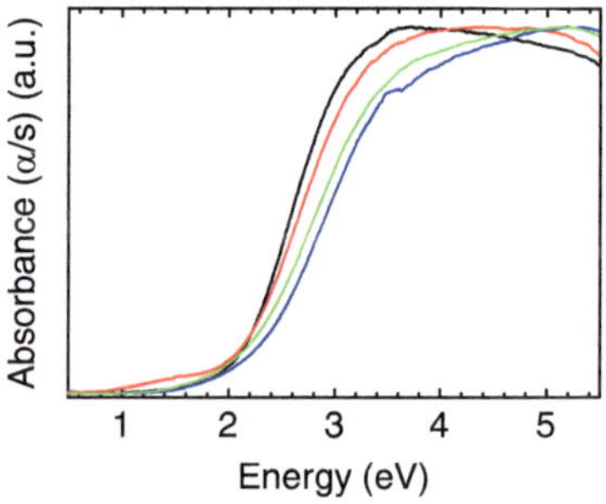

Fig. 13 Optical absorption spectra of mesoporous NU-GeSi-1 (*black*), NU-GeSi-2 (*red*), NU-GeSi-3 (*green*) and NU-GeSi-4 (*blue line*) semiconductors

strongly depends on the chemical composition and systematically increases as the Si fraction in the walls increases. Analogous blue-shift in energy band gap of a nanoporous Ge/Si alloy semiconductor relative to pure mesoporous germanium has been observed by Sun et al. [44].

4 Mesoporous Metal/Ge-based Semiconductors

Because the (Ge_9)-clusters can act as ligands to metal ions, a large variety of polymeric assemblies can form by coordination to metal ions. Indeed, the coordinative framework assembly of coupling of $[Ge_9]^{4-}$ with various linking metal ions (i.e., In, Sb, Sn, Pb, Cd) lends itself to producing a wide variety of mesoporous binary compound semiconductors [55]. These systems can be viewed as porous intermetallics, a class of materials that currently is undeveloped. As described above, templated growth occurs through the assistance of structure-directing EDMHEAB surfactant in a mixture of formamide/ethylenediamine solution. Mesoporous NU-MGe-2 structures are obtained after removal of the surfactant inside the pores by an ion-exchange procedure with ammonium nitrate following mild heating under vacuum.

$$\mathrm{EDMHEAB} + [\mathrm{Ge_9}]^{4-} + (4/\mathrm{n})\mathrm{M^{n+}} \rightarrow (\mathrm{EDMHEA})_x[\mathrm{M_{4/n}Ge_9}]$$
$$(\mathrm{EDMHEA})_x[\mathrm{M_{4/n}Ge_9}] + \mathrm{NH_4}^+ \rightarrow (\mathrm{NH_4}^+)_x[\mathrm{M_{4/n}Ge_9}] \rightarrow \underset{(\text{NU-MGe-2})}{\mathrm{H_xM_{4/n}Ge_9}} + \mathrm{NH_3}$$
$$(\mathrm{M} = \mathrm{In, Sb, Sn, Pb, and\ Cd}) \qquad (4)$$

The NU-MGe-2 materials possess well-ordered pore structures in hexagonal symmetry. Figure 14 displays typical TEM images and the corresponding fast-Fourier transforms obtained from mesoporous NU-MGe-2. The TEM images show a mesoscopic order of discernible domains of hexagonally arranged cylindrical pore channels in *p6mm* symmetry, analogous to MCM-41 [56].

The N_2 physisorption experiments on mesoporous NU-MGe-2 show typical type-IV adsorption branches with a distinct condensation step at relative pressure (P/P_o) ~0.16, suggesting well-defined mesopores. These materials have porous structure with BET surface areas in the range of 127–277 m^2/g, pore volumes in the range of 0.15–0.26 cm^3/g, and BJH pore sizes in the 2.7–2.8 nm range. These surface areas are reasonable if we consider the heavier inorganic frameworks and correspond to silica equivalent surface areas of 403–858 m^2/g. The framework wall thickness was found to be ~2 nm for mesoporous NU-MGe-2 (M = Sb, In, Sn, and Cd) and ~2.4 nm for NU-PbGe-2, which is consistent with the larger diameter of the incorporated Pb atoms.

The oxidation state of Ge atoms and linking metal atoms in the framework structure was probed by X-ray photoelectron spectroscopy (Fig. 15). The XPS spectra suggest that the Ge and linking metal atoms are in relatively low oxidation state and have electronic configuration similar to those observed in metal-alloy phases [57].

These mesoporous intermetallic MGe_9-based structures are amorphous semiconductors as they exhibit a well-defined band gap structure associated with energy gaps in the range of 1.48–1.70 eV. The energy band gap in these materials systematically varies with the linking metal atom. The considerable widening of the energy band gap in going from bulk Ge to mesoporous metal/Ge_9 framework is similar to the pure mesoporous Ge materials discussed above, and it is caused by the substantial dimensional reduction of the framework wall thickness (~2 nm) in NU-MGe-2 [58–61].

The mesoporous structures of NU-MGe-2 are strongly photoluminescent at room temperature when excited with photons with energy above the band gap. The peak maximum of the photoluminescence (PL) spectrum varied from 740 to 845 nm, depending on the chemical composition of the framework. Interestingly, the photoemission spectra can be sensitive to organic molecules that can act as acceptors. The NU-InGe-2 was found to be active toward organic molecules, small enough to enter the pores, and engage in electron-acceptor interactions with the semiconductor framework.

For example, Fig. 16 shows the room temperature photoemission of mesoporous NU-InGe-2 upon adsorbing different amounts of tetracyanoethylene (TCNE) under excitation light at 510 nm. These results indicate that the PL emission signal

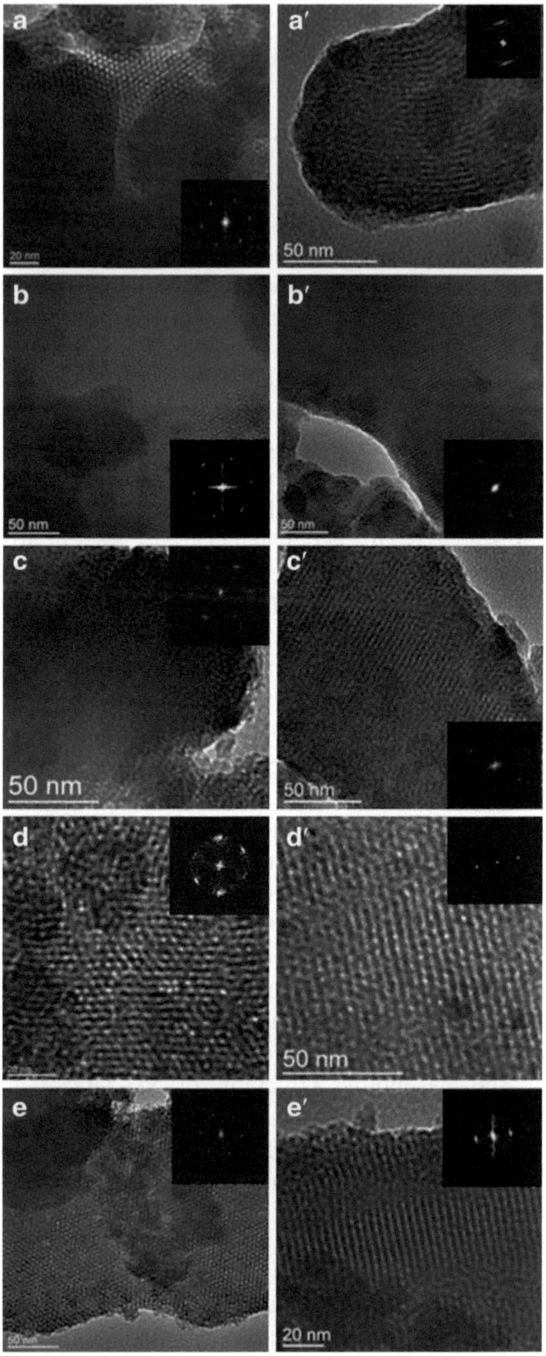

Fig. 14 TEM images of mesoporous samples of (**a** and **a**′) NU-SbGe-2, (**b** and **b**′) NU-InGe-2, (**c** and **c**′) NU-SnGe-2 (**d** and **d**′) NU-PdGe-2 and (**e** and **e**′) NU-CdGe-2 samples viewed along the pore channel axis (*left*) and perpendicular to the channel axis (*right*). The corresponding fast-Fourier transform images are shown in the *insets*. Reproduced with permission from [55]. Copyright 2008 American Chemical Society

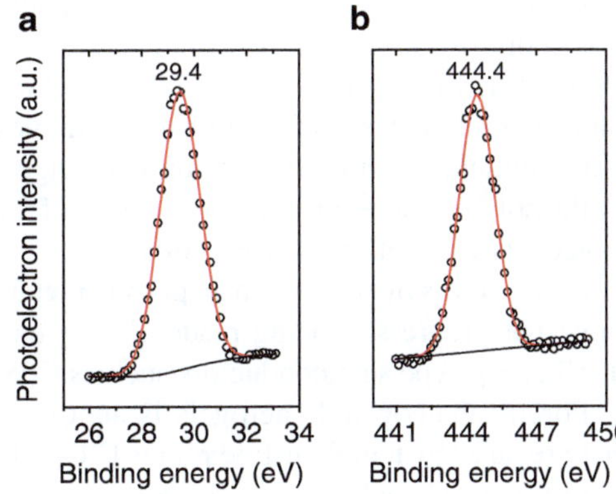

Fig. 15 Typical high-resolution (**a**) Ge 3d and (**b**) In 3d5/2 core-level photoelectron spectra of mesoporous NU-InGe-2

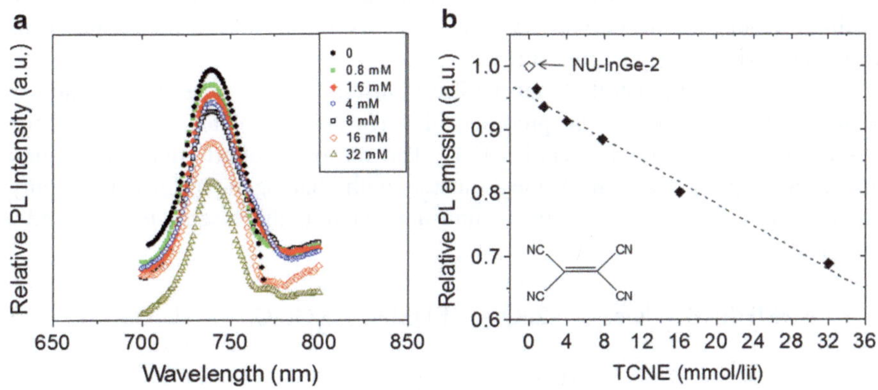

Fig. 16 (**a**) Relative PL emission spectra of mesoporous NU-InGe-2 and TCNE@NU-InGe-2 semiconductors as a function of the TCNE concentration, and (**b**) the TCNE concentration dependence of the PL emission peak of mesoporous NU-InGe-2. The straight line fitting the data is given as a guide to the eye ($R^2 = 0.9938$). Reproduced with permission from [55]. Copyright 2008 American Chemical Society

of NU-InGe-2 is suppressed via energy transfer to the electron-acceptor TCNE molecules. However, because of the absence of analogous energy transfer, the PL remains unchanged when the materials is exposed to electron-donor species such as tetrathiafulvalene (TTF) molecules.

5 Germanium-Based Chalcogenide Materials

The early attempts to produce mesostructured metal chalcogenide involved surfactant-templated cross-linking polymerization of Zintl clusters (MQ_4, M_4Q_{10} and Sn_2Q_6; M = Ge, Sn; Q = S, Se, Te) with various metal ions (Hg^{2+}, Cd^{2+}, Pt^{2+}, Zn^{2+}, In^{3+}, Ga^{3+}, Sb^{3+}, Bi^{3+}, Sn^{2+}, Sn^{4+}) [45, 62–67]. A family of mesostructured materials with a wide range of chemical compositions and pore architectures, from wormholes-to-hexagonal-to-cubic, can be prepared upon changing the chalcogenide cluster, linking metal ion, and surfactant chain length. These mesostructured chalcogenide materials feature a controllable and regular pore structure with interesting optical, electronic, and ion-exchange properties. However, these materials could not be rendered porous because of the inability to remove the surfactant molecules without framework collapse.

So far, the synthesis of mesoporous metal chalcogenides remains an open challenge but progress is being made. A few examples have been reported, including II–VI group-type semiconductors such as CdS [68, 69], ZnS [70], and CdTe [17].

Porous chalcogenide aerogels is another broad class of non-oxidic framework that prepared by template-free routes [71–73]. These materials possess a continuous nanostructured chalcogenide framework that is penetrated by a random network of nanopore channels. Because these high surface area structures are random and not exhibit long-range pore periodicity, such systems are outside of the scope of this review and will not be covered further.

Recently, we demonstrated that the Zintl clusters $[Ge_9]^{4-}$ react with chalcogen atoms (S, Se, and Te) in the presence of surfactant templates to form ordered mesoporous Ge-rich chalcogenides [74]. The mesostructured frameworks grow through a coupling reaction of (Ge_9)-clusters with chalcogens in formamide/ethylenediamine mixture solution in an unusual reaction that seems to be a redox process (5).

$$\text{EDMHEAB} + [Ge_9]^{4-} + 2n\text{Q} \rightarrow (\text{EDMHEA})_x Ge_9 Q_n \rightarrow \text{NU-GeQ-}N \qquad (5)$$

$$(\text{Q} = \text{S, Se, Te}; n = 1, 2)$$

The NU-GeQ-1 and NU-GeQ-2 materials possess a Ge-rich chalcogenide framework perforated by a regular array of pore channels in hexagonal and cubic symmetry. Figure 17 displays typical TEM images and the corresponding fast-Fourier transform patterns of mesoporous NU-GeQ-1 and NU-Ge-2. According to XRD patterns and TEM images, the NU-GeQ-1 and NU-GeS-2 feature porous structure of 2D hexagonal arrangement of parallel pore channels, while the NU-GeSe-2 and NU-GeTe-2 possess regular arrays of mesopores assigned to body-center cubic *Ia*-3*d* symmetry. These materials contain anionic "Ge_xQ" networks which are charge-balanced with the cationic surfactant molecules. An open question is the bonding mode between the clusters and the chalcogen atoms. In the absence of a crystal structure, this will be difficult to address, and also such Ge-rich chalcogen compositions are not known in the literature even for dense crystalline phases. The stability of NU-GeQ-1 and NU-GeQ-2, however, suggests that such compositions may be crystallizable if the appropriate counterions were used.

The mesoporous NU-GeQ-*N* chalcogenides exhibit high internal surface areas in the range of 350–516 m^2/g and quite narrow size distribution of pores with pore diameter at ~2.9–3.1 nm. The framework wall thickness varies from ~1.5 to

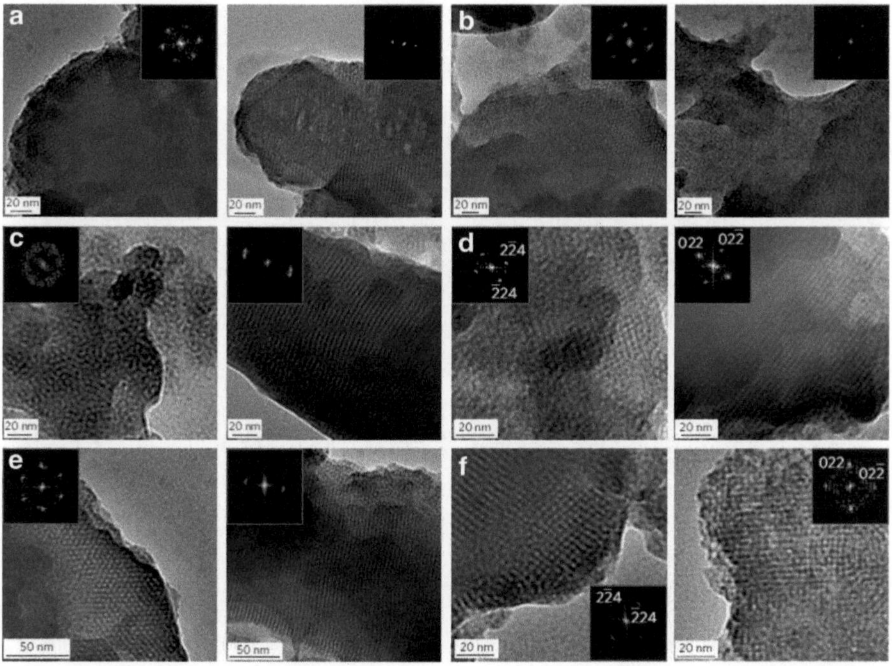

Fig. 17 TEM images of hexagonal mesoporous (**a**) NU-GeS-1, (**b**) NU-GeS-2, (**c**) NU-GeSe-1, and (**d**) NU-GeTe-1 viewed along the pore channel axis (*left images* in each part) and perpendicular to the pore channel axis (*right image* in each part) and TEM images of cubic mesoporous (**e**) NU-GeSe-2 and (**f**) NU-GeTe-2 taken along the [110] direction (*left image* in each part) and along the [100] direction (*right image* in each part). *Insets* in each panel show the corresponding FFT images. Reproduced with permission from [74]. Copyright 2009 Nature Publishing Group

~2.0 nm in going respectively from sulfur to tellurium analogs, which is consistent with the increasing atomic diameter.

These structures feature a unique functionality that is the high polarizable internal pore surface. The surface polarizability arises from the "soft" chalcogen atoms that comprise the framework. The heavier chalcogen atoms are more polarizable than, for example, oxygen or silicon atoms. The polarizability trend is as follows: O < S < Se < Te and Si < Ge. Therefore, the surface polarizability of the Ge_xQ-derived materials is expected to be higher than those of metal oxides, porous carbons, and organic polymers. Because of this, it is expected that the strength of interactions of the various gases with the porous surface will be dependent on their polarizability. The more polarizable molecules will interact more strongly. As a result, the NU-GeQ-1/2 and NU-PbGeTe-1 structures show promise for developing gas mixture separation processes based on surface polarizability, rather than size and shape of the pore. Figure 18 shows the adsorption isotherms for CO_2 and H_2 and gas mixture separation results by passing $H_2/CH_4/CO_2/CO$ gases through mesoporous NU-GeTe-1 and NU-GePbTe-1 materials. These experiments clearly support excellent solubility separation of carbon dioxide

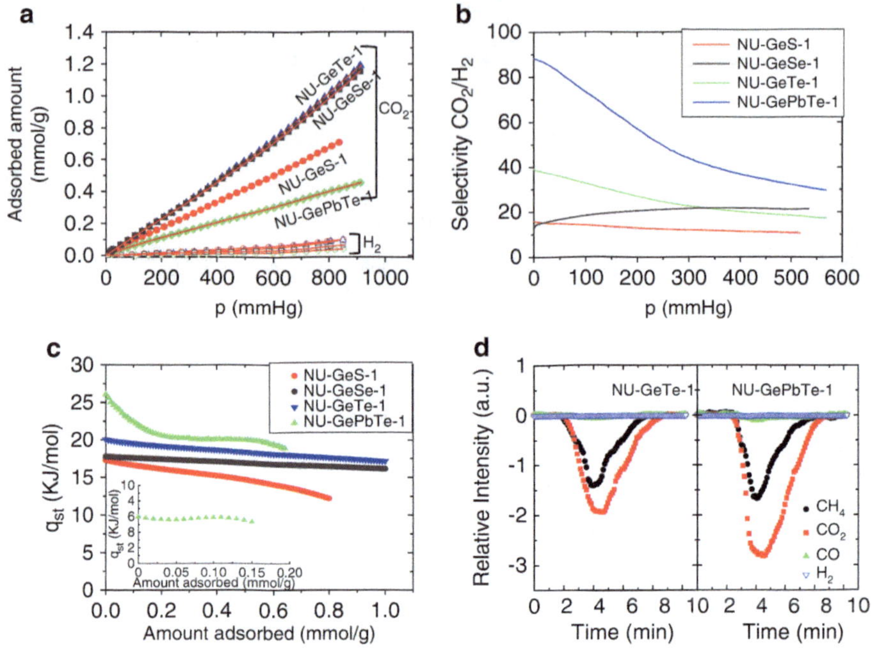

Fig. 18 (**a**) Adsorption isotherms for CO_2 (*filled symbols*) and H_2 (*open symbols*) of mesoporous NU-GeS-1 (*black*), NU-GeSe-1 (*red*), NU-GeTe-1 (*blue*) and NU-GePbTe-1 (*green symbols*) at 0°C. The *red lines* are fits to the data. (**b**) The CO_2/H_2 selectivities predicted by applying the ideal adsorption solution theory to the CO_2 and H_2 single-component isotherms at 0°C of NU-GeQ-1 and NU-PdGeTe-1 samples. (**c**) The loading dependence of isosteric heat of the CO_2 adsorption that is determined by fitting the isotherms at −10 and 0°C to appropriate virial-type equations. *Inset*: The loading dependence of isosteric heat of adsorption for H_2 of the NU-GePbTe-1 sample. (**d**) The gas mixture separation curves ($H_2/CH_4/CO_2/CO$ 20:22:18:5 v/v in N_2) from NU-GeTe-1 and NU-GePbTe-1 solids at 0°C. Reproduced with permission from [74]. Copyright 2009 Nature Publishing Group

and methane over hydrogen. Indeed, the separation performance of these porous chalcogenides seems to be related to the polarizability strength of probe molecules and following the range $CO_2 > CH_4 > CO > H_2$.

6 Future Prospects

The paradigm of how molecular-based Zintl compounds can be used to make porous materials has been demonstrated. The potential of Zintl phases to act as starting materials for the synthesis of a wide variety of open frameworks is enormous because of the broad diversity of such phases. Not only Zintl anions such as E_9^{4-} and E_4^{4-} (E = Ge, Si, Sn, Pb), As_7^{3-}, Bi_4^{2-}, Sb_7^{3-}, and In_{11}^{7-} composed of single elements but also binary and ternary ones, such as $Sn_xE_{9-x}^{4-}$ (E = Ge, Pb),

$Pb_2Sb_2^{2-}$, $Sn_2Bi_2^{2-}$, ESn_8^{3-} (E = Sb, Tl), $In_4Bi_5^{3-}$, and $Zn_9Bi_{11}^{5-}$, can be used in imaginative ways [75–79].

The materials described in this chapter have amorphous frameworks, and therefore in many respects they are still limited especially with regard to electronic transport. A future challenge is the realization of crystalline frameworks with mesoporous properties. The achievement of crystalline versions made from Zintl building blocks will represent a big step forward which will allow useful doping and higher carrier mobilities to be observed. This will lead to new types of potential applications.

Acknowledgments These studies were supported primarily by the Nanoscale Science and Engineering Initiative of the National Science Foundation under NSF Award Number EEC-0647560. This work made use of the J.B. Cohen X-ray Diffraction facility and the Electron Probe Instrumentation Center (EPIC) and Keck Interdisciplinary Surface Science (Keck-II) facility of NUANCE center at Northwestern University.

References

1. Ozin GA (1995) Adv Chem Ser 245:335
2. Weiss D, Richter K, Menschig A (1993) Phys Rev Lett 70:4118
3. Turton R (1995) The quantum dot. Oxford University Press, Oxford, UK
4. Ruda HE (1992) Science 283:646
5. Corma A, Atienzar P, Garcia H et al (2004) Nat Mater 3:394
6. Alivisatos AP (1996) Science 271:933
7. Wan Y, Yang H, Zhao D (2006) Acc Chem Res 39:423
8. Soler-Illia GJ, Sanchez C, Lebeau B et al (2002) Chem Rev 102:4093
9. Kanatzidis MG (2007) Adv Mater 19:1165
10. MacLachlan MJ, Coombs N, Ozin GA (1999) Nature 397:681
11. MacLachlan MJ, Coombs N, Bedard RL et al (1999) J Am Chem Soc 121:12005
12. Trikalitis PN, Rangan KK, Bakas T et al (2001) Nature 410:671
13. Trikalitis PN, Kanatzidis MG (2005) J Am Chem Soc 127:3910
14. Riley AE, Tolbert SH (2003) J Am Chem Soc 125:4551
15. Rangan KK, Trikalitis PN, Kanatzidis MG (2000) J Am Chem Soc 122:10230
16. Trikalitis PN, Rangan KK, Kanatzidis MG (2002) J Am Chem Soc 124:2604
17. Jiang T, Ozin GA (1998) J Mater Chem 8:1099
18. Froba M, Oberender N (1997) Chem Commun 1729
19. Yaghi OM, Sun Z, Richardson DA et al (1994) J Am Chem Soc 116:807
20. Kim K-W, Kanatzidis MG (1992) J Am Chem Soc 114:4878
21. Kim K-W, Kanatzidis MG (1998) J Am Chem Soc 120:8124
22. Loose A, Sheldrick WS (1997) Z Naturforsch B 52:687
23. Ding N, Kanatzidis MG (2006) Angew Chem Int Ed 45:1397
24. Li HL, Laine A, O'Keeffe M et al (1999) Science 283:1145
25. Jiang T, Ozin GA, Bedard RL (1994) Adv Mater 6:860
26. Manos MJ, Iyer RG, Quarez E et al (2005) Angew Chem Int Ed 44:3552
27. Manos MJ, Chrissafis K, Kanatzidis MG (2006) J Am Chem Soc 128:8875
28. Guloy AM, Ramlau R, Tang Z et al (2006) Nature 443:320
29. Ammar A et al (2004) Solid State Sci 6:393
30. Dunlap WC (1950) Science 122:419
31. Bundy FP, Kasper JS (1963) Science 139:340

32. Tauc J, Grigorov R, Vancu A (1966) Phys Status Solidi b 15:627
33. Riley AE, Korlann SD, Richman E et al (2006) Angew Chem Int Ed 45:235
34. Kesanli B, Fettinger J, Gardner DR et al (2002) J Am Chem Soc 124:4779
35. Coicoechea JM, Sevov SC (2005) Angew Chem Int Ed 44:4026
36. Esenturk EN, Fettinger J, Lam YF et al (2004) Angew Chem Int Ed 43:2132
37. Eichhorn BW, Haushalter RC, Pennington WT (1988) J Am Chem Soc 110:8704
38. Kesanli B, Fettinger J, Eichhorn BW (2001) Chem Eur J 7:5277
39. Ugrinov A, Sevov SC (2004) Chem Eur J 10:3727
40. Armatas GS, Kanatzidis MG (2006) Nature 441:1122
41. Billing SJL, Kanatzidis MG (2004) Chem Commun 7:749
42. Nakamura Y, Watanabe K, Fukuzawa Y et al (2005) Appl Phys Lett 87:133119
43. Armatas GS, Kanatzidis MG (2006) Science 331:817
44. Sun D, Riley AE, Cadby AJ et al (2006) Nature 441:1126
45. Trikalitis PN, Ding N, Malliakas C et al (2004) J Am Chem Soc 126:15326
46. Jacobs DJ, Thorpe MF (1995) Phys Rev Lett 75:4051
47. Klimov VI et al (2000) Science 290:314
48. Armatas GS, Kanatzidis MG (2008) Adv Mater 20:546
49. Barrett EP, Joyner LG, Halenda PH (1951) J Am Chem Soc 73:373
50. Kruk M, Jaroniec A, Sayari A (1997) J Phys Chem B 101:583
51. Lu X, Korgel BA, Johnston KP (2005) Chem Mater 17:6479
52. Hollinger G, Kumurdjian P, Mackowski MJ et al (1974) J Electron Spectros Relat Phenomena 5:237
53. Chastain J, King RC Jr (1995) Handbook of X-ray photoelectron spectroscopy: a reference book of standard spectra for identification and interpretation of XPD data. Physical Electronics, Eden Prairie
54. Freeman EC, Paul W (1979) Phys Rev B 20:716
55. Armatas GS, Kanatzidis MG (2008) J Am Chem Soc 130:11430
56. Kresge CT, Leonowicz ME, Roth WJ et al (1992) Nature 359:710
57. Guloy AM, Corbett JD (1996) Inorg Chem 35:4669
58. Nakamura Y, Watanabe K, Fukuzawa Y et al (2005) Appl Phys Lett 87:1
59. Lu X, Zigler KJ, Ghezelbash A et al (2004) Nano Lett 4:969
60. Taylor BR, Kauzlarich SM, Lee HW et al (1998) Chem Mater 10:22
61. Yang H, Yao X, Wang X et al (2003) J Phys Chem B 107:13319
62. Wachhold M, Rangan KK, Bilinge SJL et al (2000) Adv Mater 12:85
63. Wachhold M, Rangan KK, Lei M et al (2000) J Solid State Chem 152:21
64. Rangan KK, Billing SJL, Petkov V et al (1999) Chem Mater 11:2629
65. Rangan KK, Trikalitis PN, Bakas T et al (2001) Chem Commun 809
66. Trikalitis PN, Bakas T, Papaefthymiou V et al (2000) Angew Chem Int Ed 39:4558
67. Korlann SD, Riley AE, Kirsch BL et al (2005) J Am Chem Soc 127:12516
68. Braun PV, Osenar P, Stupp SI (1996) Nature 380:325
69. Gao F, Lu Q, Zhao D (2003) Adv Mater 15:739
70. Li J, Kessler H, Soulard M et al (1998) Adv Mater 10:946
71. Mohanan JL, Arachchige IU, Brock SL (2005) Science 307:397
72. Bag S, Trikalitis PN, Chupas PJ et al (2007) Science 317:490
73. Bag S, Gaudette AF, Bussell ME et al (2009) Nat Chem 1:217
74. Armatas GS, Kanatzidis MG (2009) Nat Mater 8:217
75. Corbett JD (1985) Chem Rev 85:383
76. Xu L, Sevov SC (2000) Inorg Chem 39:5383
77. Coicoechea JM, Sevov SC (2006) Angew Chem Int Ed 45:5147
78. Fässler TF (2001) Coord Chem Rev 215:347
79. Sevov SC (2002) Zintl phases. In: Westbrook JH, Fleischer RL (eds) Intermetallic compounds: principles and practice, vol 3. Wiley, Chichester, UK

Index

MIX
Papier aus verantwortungsvollen Quellen
Paper from responsible sources
FSC® C105338

If you have any concerns about our products,
you can contact us on
ProductSafety@springernature.com

In case Publisher is established outside the EU,
the EU authorized representative is:
Springer Nature Customer Service Center GmbH
Europaplatz 3, 69115 Heidelberg, Germany

Printed by Libri Plureos GmbH
in Hamburg, Germany